THE TINY BUT MIGHTY Farm

Cultivating High Yields, Community, and Self-Sufficiency from a Home Farm

Jill Ragan

OF *Whispering Willow Farm*

COOL
SPRINGS
PRESS

Inspiring | Educating | Creating | Entertaining

Brimming with creative inspiration, how-to projects, and useful information to enrich your everyday life, quarto.com is a favorite destination for those pursuing their interests and passions.

26 25 24 23 22   1 2 3 4 5

ISBN: 978-0-7603-7645-4

Digital edition published in 2023

eISBN: 978-0-7603-7646-1

Library of Congress Cataloging-in-Publication Data

Names: Ragan, Jill, author.

Title: The tiny but mighty farm : cultivating high yields, community, and self-sufficiency from a home farm / Jill Ragan.

Other titles: Cultivating high yields, community, and self-sufficiency from a home farm

Description: Beverly, MA : Cool Springs Press, 2023. | Includes index. | Summary: "Discover how to turn a typical suburban property into a food-growing machine in The Tiny But Mighty Farm. Feed your family and maybe even turn a profit"— Provided by publisher.

Identifiers: LCCN 2022033861 | ISBN 9780760376454 (trade paperback) | ISBN 9780760376461 (ebook)

Subjects: LCSH: Gardening. | Vegetable gardening. | Urban agriculture. | Handbooks and manuals.

Classification: LCC SB320.9 .R34 2023 | DDC 635—dc23/eng/20220822

LC record available at https://lccn.loc.gov/2022033861

Design and Page Layout: Amy Sly
Photography: Bailee McKay Photography
Illustration: Abigail Diamond

Printed in USA

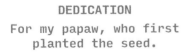

DEDICATION

For my papaw, who first
planted the seed.

Contents

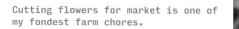
Cutting flowers for market is one of my fondest farm chores.

Introduction

The Tiny Farm Journey

Hi friend,

I'm Jill Ragan, Garden Grower at Whispering Willow Farms nestled in the hill country of Central Arkansas. I've been on this garden-growing adventure for the last 10-plus years. I never finished my agriculture degree, nor do I come from a prominent family of well-to-do farmers, but I do come from a long, honest line of tried-and-true backyard gardeners who grew food out of necessity and gardened with passion.

My grandfather, who I will affectionately refer to as Papaw throughout this book, was my first and favorite mentor. Most of my summers were spent with him seeding, weeding, watering, and harvesting. My grandmother would also recruit me to help can jams and preserve vegetables to last through the winter months. The wisdom they sewed into me are not lessons you can learn from a textbook or a quick google search. They are core values and principles that guide you as a farmer.

Watching them work the land as a way of life piqued my curiosity and planted the seeds of desire to one day grow my own food. As I said, I have spent the last 10 years of my adult life dedicating my time, energy, and resources to learning this skill and growing my craft as a farmer.

My husband and I now have a small farm where we grow food for our family and our community and teach others how to plan and steward their own farms and dreams. This is more than just a career for me; it's my purpose.

We've all been called to something. For me, it's the garden. It's raising a generation that's aware of their food and how it's grown and cared for. It's to encourage every single person who turns the pages of this book that they were called to do great

I see each and every one of you working hard and putting in the sweat required to make your gardens and farms successful. If no one has told you how adequate you are to do this job, I am here to tell you. You are brave, capable, and destined to grow beautiful food.

things. That, you, yes you, can grow abundant food and learn and grow as a gardener. I've never been so proud of the work I've done before and I hope throughout this book you will feel electrified with purpose, desire, and the grit to dig your heels in and feel this same emotion about your farm and garden too.

Before we dive into best practices, how-to's, and special techniques, I want to start from the beginning and take you back to where it all began for me. I believe that the process of gardening and the lessons it teaches us are just as important as harvesting food.

I can recall my childhood summers vividly, there were buckets toppling over full of juicy blueberries, and blackberries picked fresh that morning from the garden. There'd be tomatoes stacked atop one another on the kitchen counter, calling our name as summer's tastiest treat and a reward for pulling weeds. This is my memory as a kid of summers spent at Papaw's.

But where did such delicious abundance come from? It came from my Papaw's humble backyard garden. He did not need a 500-acre (202 ha) farm to put delicious food on the table. He showed us just the opposite. He instilled in us the value that dreaming big meant setting attainable, achievable goals. He taught us the value of a small garden—that even a small garden can have a significant impact on the fruit it produces, on our communities, and on our hearts. He showed me in a tangible way how 10 blueberry bushes of suitable varieties could create enough to provide our family with jelly to last a year.

Often when we think of high yields, we think of large farms. Which, yes, can be the case. But, my Papaw taught me that if done correctly, small-scale gardening could produce more than we needed.

Later in life, I picked up this love of gardening that was instilled in me as a child and ran wild with it. I had my first garden nestled smack dab in a subdivision in the middle of town in my tiny backyard. Brace yourselves. It was a big one— roughly 150 square feet (14 sq. m) of freshly tilled ground waiting for pockets full of seeds to make their way into it. I knew nothing back then.

I planted way more than my little space could handle, choked out my tomatoes with out-of-control watermelon plants, and shed actual tears when I harvested that first melon—the one that ended up rotting on the kitchen counter. It turns out that growing things you plan on eating plays a significant role in the success of your garden. While it was beautifully chaotic, I learned some practical things from my first garden that I would take to my next home.

This time I made my way into a cute log cabin nestled on an acre (.4 ha) of land. I had 400 square feet (37 sq. m) of in-ground growing space and around ten 4 x 4 foot (1.2 x 1.2 m) raised beds. I was doing this growing thing and making Papaw proud.

I invested in the right varieties (maybe too many, if I'm honest), started seeds, and watched them grow. I was determined to see just how much that little area could produce; I pushed the limits,

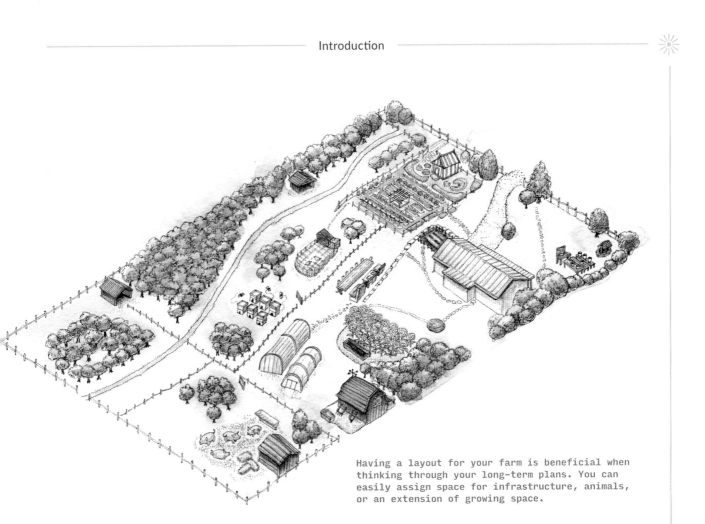

Having a layout for your farm is beneficial when thinking through your long-term plans. You can easily assign space for infrastructure, animals, or an extension of growing space.

and abundance sprung forth. I produced so much food in that tiny space.

Do you desire to grow a delicious bounty in a small space too? There are so many ways to make your tiny but mighty farm thrive. Together, we'll explore your growing options, select the right varieties for high yields, steward the soil, consider helpful tools & structures, and, best of all, navigate small spaces to have a significant impact on your family and community.

This book is not meant to be a coffee table accessory or a pretty decor item. This book is intentionally filled with questions where you have space to jot down your answers. I have created useful charts and graphs you can reference while you're out in your garden nurturing your soil, sowing the seeds, and harvesting.

My heartfelt vision for this book is that it would provide a roadmap of knowledge and inspiration as you continue pursuing a sustainable lifestyle through growing your own food.

One of the pillars of my farm's development is that community is at the core of what we do. I see this book as an extension of our community.

Thank you for being a part of the Whispering Willow Farm community. And thank you for doing your part to be a good steward of the earth and to grow beautiful food for your family and friends.

I'll talk to you soon,

Jill

Mexican sour gherkins (aka cucamelons) are one of m
kids' favorite summertime snacks. They are a cross
between a cucumber and a watermelon.

CHAPTER ONE

Small-Farm Values

The future of agriculture will come from people, not the technology—from a new generation of farmers who embrace small-scale, ecological, and nourishing farming techniques.

—Jean-Martin Fortier

Many of us have different views of what it looks like to have a farm. For some, that looks like 1,000 acres (400 ha) covered in perfectly ordered rows. For my Papaw, that looked like gathering neighborhood friends and family to share a tiller and grow a communal garden. It was less of a *want to* and more of a *have to*, but those hard-earned skills and principles he developed have now been passed down through generations, and while I may choose to practice growing food differently than he did, this book is very much the fruits of his hard labor.

For me, growing food means walking through my garden gate and seeing beans dangling from an arched trellis or kale and rainbow chard protruding from the sides of the bed—bees buzzing around pollinating the basil and mint.

CULTIVATING FAMILY

Purple-podded pole beans have been a
staple in our house for years. They freeze
and can exceptionally well.

Growing food also means friends and family gathered around the table cutting into freshly baked sourdough bread and nearby are cherry tomatoes sprinkled on top of a freshly picked salad. Community gatherings to support one another's dreams and visions are what growing food means to me. And we can't forget about the dahlias blooming in the cottage garden, or 'Kellogg's Breakfast' ripe tomatoes hanging on the vine. Farming sparks a joy that is so abundant that it forces me to connect to the people and beauty around me daily. For others, growing food simply means growing enough to sustain their families, and some envision small market gardens, hoping to feed their immediate communities.

WHAT IS A MARKET GARDEN?

Market gardens are a larger-scale production of vegetables, fruits, and flowers for sale within the surrounding community. Though its scale is larger than a home garden, many of the same principles apply.

Are you in the beginning stages of creating your small garden and hoping to cut food costs? Are you selling at a farmer's market to bring in extra income? Would you consider yourself a beginner gardener just wanting to grow food or an experienced gardener wanting to scale to the market grower?

Whatever scale you choose to garden and farm, there is a place at the table. Sustainability isn't tied to scale and vice versa. Both large- and small-scale farms can operate in a manner that heals or harms our environment.

Generally speaking, one by one, small farms are reconnecting the pieces to sustainability and connecting people to their local food systems. There are 2.5 billion people in this world that live on small farms. That is a lot. Many of them

The Bell Urban Farm plant sale every year is such a wonderful representation of community and supporting small farmers.

work closely with the land and invest in their local environment. They protect biodiversity, use livestock to prepare their fields, and practice growing techniques that create healthier soils for growing healthier plants—while preserving the pollinators and ecosystems from harsh chemicals and pesticides.

The concept can be pretty straightforward. What are you trying to accomplish?

Small-scale growers who implement good stewardship practices on any level show their community and families in a very tangible way how they care for the food that ends up on their tables. The idea shouldn't be to grow food; we should all adopt the concept of growing food in a sustainable way that betters the environment and teaches principles that many can adapt and quickly implement themselves.

I remember volunteering on a few small urban farms when I pursued getting my own farm. I learned a lot about growing food and how it directly impacted people. Many of the places I visited had more than a goal of producing food for themselves or the community. Instead, they also had plans to extend an invitation to aspiring farmers and gardeners like myself who could one day contribute to the vision of feeding our communities together. They understood that

we are better together than we are apart. That's something I will always carry with me and choose to implement on my tiny farm—the pursuit of the community over competition and sustainability over scale.

No matter the scale, small farms and gardens play a vast role in the betterment of our communities.

I often joke that no one leaves my house empty-handed, but that statement couldn't be more accurate. Our communities are an extension of who we are. We grow food for sustainability; we succeed because we are called to our work. Still, ultimately, we grow because we are knitting together the very intricate thing that sustains us all.

Without stewardship of the land, there is no food, and without the stewardship of the community, there is no need to grow. It's all interwoven. One doesn't happen without the other.

Without stewardship of the land, there is no food, and without the stewardship of the community, there is no need to grow.

Let's cultivate food to impact a better tomorrow, serve our communities as though it matters, and extend grace to ourselves during the learning process of becoming a grower.

When I embarked on this journey, I knew it would be challenging, but I felt hopeful that the challenge would pay a hefty reward, and I'm not talking about money here. The relationships I worked so hard on cultivating are the driving factors that keep our small farm going. Funnily enough, when I first started, I thought we would be the ones making an impact. It turns out I had it all wrong.

Our farm sits on only 4.3 acres (1.7 ha), but this is more than enough space because we chose suitable land and established good plans. We spent the first 6 months after we moved onto our property emphasizing the importance of soil health. We analyzed our soil and with the results immediately worked toward adding the correct nutrients back into our land for a more successful yield later in the season. I also took time to

establish crop plans for our property and focused on succession planting—more than I had done in the past—to ensure we would have continuous food throughout all the seasons. Knowing what you want to plant, and how to continue to grow those successions makes a world of difference, and throughout the succession planting, I am adding nutrients back into the soil. It really is a win-win.

Our small-farm values don't lie in the number of acres we own. They lie within what we can pour back into it—adding to and not taking away. Healthy soil is the foundation of our food system.

Small farms have many functions, from preserving soil quality to preventing land degradation and benefiting society and the biosphere.

Many larger farms tend to plant monocultures because they are the easiest to manage with large, heavy machinery. However, most small farms produce various crops and diversify their crop rotations while working in harmony with nature. A space that would otherwise grow weeds is now occupied with multiple crops. Many small farms rotate their crops and livestock—continuing to build their soil. So, while the large farm will likely always produce more yield, the intensive small farms will have fewer outputs and have a higher yield per area.

WHAT IS A MONOCULTURE?
Monoculture is the practice of growing a single crop species in a field at a time.

Most large, conventional farms grow things that can be stored and traded, such as grain and beans. They also cover a lot of land, using only a few people to work the land. They depend primarily on capital to bankroll the farm, and rely on machinery, chemicals, and material inputs.

On the contrary, small intensive farms utilize inputs created through people. Their land is cultivated, harvested, packed, and delivered by human hands. By doing this, the small farm is

'Napoli' carrots are the perfect size for fresh eating, canning, or, my favorite, fermenting.

Our second round of
summer flowers are
being seeded in the
greenhouse.

creating meaningful work for the grower and the people partnered with them. I believe people should be viewed as an asset to any small-farm value.

> **The ultimate goal of farming is not the growing of crops, but the cultivation and perfection of human beings.**
> —Masanobu Fukuoka

Small-scale farms promote community. There is something special about walking into a mom-and-pop farm stand that sells beautiful food that was grown locally by a farmer that the community knows and trusts. As consumers, we invest in our community's economy when we shop locally. Doing this we support businesses and other striving farmers. We are also taking away the corporate label when we grow it ourselves. The guesswork about the produce is no longer an issue. Were chemicals used to produce this? How far was it transported to get to me? Questions like these are no longer in play because we were a part of

the entire process of growing, harvesting, and preserving our food.

Large farms are capital intensive, while small-scale farming and gardening is different. It is undoubtedly more labor-intensive, but it is meaningful and beautiful work, and its impact and benefits on the gardener and the whole Earth should be valued.

I realize that this book is designed to teach you how to grow overflowing yields in your small space. But I would not hold to my values, which you will see mentioned below, if I did not discuss all the actual benefits small farms have in our communities. If you read this book and never sell an ounce of your produce, this is valuable. If you read this book and intend to sell all of your produce, this is also valuable because the value lies with our becoming better and stronger together.

The reality is that small farms create numerous jobs and merge communities. They improve the health and longevity of the land and the health and longevity of the people. They also produce a more resilient foundation for the global food system.

Now that we've talked through some of the differences between large-scale and small intensive farming, I want to set you up with what you need

Routine farm walks will prove beneficial for yourself and your farm. It holds you accountabl for the chores you need to accomplish and helps you keep notes on varieties, pests, diseases, likes, and dislikes.

so you can choose the best farm for you. I also want to discuss creating the right plan if you already own your farm. Having the land and the right strategy will ensure your tiny but mighty farm will thrive.

Considerations for your tiny farm

Value and Vision

Throughout this book, you will hear me encourage you to write out your dreams and to set realistic expectations for yourself and your farm. The crops will come and go, and the style of your gardening will likely change over time, but one thing that remains will be your values. These values will be the foundation and the pillars that help you grow. That driving factor gets you out of

bed every day to pursue this thing that you love—growing food.

Having a vision for your farm is motivating and inspirational, and it allows you and anyone else working with you to focus on the common goal. When you have a clear idea, it helps you stay focused on the task at hand and helps you make better decisions for yourself and your farm. When you have these pillars of reference, you can make sure they align with your values before moving forward on any decision.

I am a pen and paper kind of gal. I have notebooks upon notebooks scattered about every greenhouse and lined up on bookshelves. I like to have them as a point of reference. I'm going to ask you to get out your pen and paper and write these questions down. (If a computer is more your speed, by all means, carry on.)

Doing this practice made me a better gardener. (I wouldn't ask you to do something I have not already done myself.) Write down these simple questions but really take your time to answer them. Those pillars I was referring to earlier? That's what these will be—the backbone of your farm.

Our Whispering Willow Farm core values

If you stepped into my dining room, you would immediately be greeted by giant papers taped all over the walls. These pieces of paper certainly do not go with my home decor, but they are filled with declarations and truths that are crucial for sustaining my family and our farm. The writings are our core values. It is extremely important to me that we maintain our integrity and principles as we continue to grow our business and our farm.

My husband, Nathan, and I spent countless hours writing these out and deciding what values and beliefs we truly stand on. Any opportunity that arises does not get answered until we have looked over our core values to see that the outcome aligns with our vision and end goal.

These values did not come to us overnight— they are a cumulation of lessons learned, wisdom inherited, and convictions we have built our lives upon. These are principles that have impacted us and will impact others—especially, those in our home. If I cannot express our values fluently to my own children, why am I trying to influence others?

> **Never let it be said that to dream is a waste of one's time, for dreams are our realities in waiting. In dreams, we plant the seeds of our future.**
> —Author Unknown

Our principles and values start in our homes, with our children. They're cultivated in our daily lives and in how we treat others and our environment. I do not want to be known for growing the biggest garden or having the most

Having your values where you can see them and reference them often is vital. Come back to them on the hard days, reference them for encouragement, and always evaluate them before making decisions on your farm.

lavish farm. I want to be remembered for making a substantial impact for the betterment of our society and for inviting others to join me as we place our core values at the center of what we do.

Here's your task. Answer the following questions about the **pillars**, **values**, and **vision** for your farm. Pillars are the immutable values, and the columns that hold together your farm. If you knock down your pillars, your farm will collapse. Pillars are the "what" and the "why" we do what we do. Your values are the things that you believe are important to the way you live and work. A vision statement is what your farm should be in the future.

- What are your farm pillars?
- What are your environmental values?
- What are your community values?
- What are your local economic values?
- What is your vision statement for your farm?

VISION STATEMENTS

A vision statement is an inspirational statement of an idealistic emotional future of a company or group. Vision describes the basic human emotion that a founder intends to be experienced by the people the organization interacts with.

Whispering Willow Farm vision and mission statement

At Whispering Willow Farm, we equip individuals inspired to grow food through educational courses, resources in our shop, and daily advice through our online community. We want to be good stewards of the Earth and see our future generations continue to shift towards sustainability.

Now, you may ask yourself different questions to find the right answer for your small-farm values, but for me, these were the things I asked myself. These were the building blocks on which my farm was rooted.

Focus on the values you've written down and give yourself the freedom to dream wildly about the vision statement. You might be surprised what you can accomplish when you let yourself dream.

Space and budget

We live in a culture where the motto "go big or go home" is ingrained in our minds. I want to challenge you to think differently. Think about maxing out your tiny space with efficient systems.

Let's start at the beginning. You cannot have a farm without the land, and it will likely be your most considerable expense unless you inherited

Nothing is more satisfying than checking off seedlings that have been started for the season.

Farm Cost Analysis*

	Less than 0.25 acre (.1 ha)	Less than 0.5 acres (.2 ha)	1 acre (.4 ha)	More than 3 acres (1.2 ha)
Seed Starting	grow lights	grow lights, 250 square feet (23 sq. m) greenhouse	500 square feet (47 sq. m) greenhouse	germination chamber, 1,000 square feet (93 sq. m) greenhouse
Layout	raised beds	permanent beds	permanent beds, blocks, or rows	permanent beds, blocks, or rows
Structures	tool shed	low tunnels, fridge	high tunnel, walk in cooler, covered pack-shed	high tunnels, walk-in cooler, enclosed pack-shed, loading dock
Infrastructure	electric, water	electric, propane heat, water	electric, propane heat, irrigation	electric, propane heat, well or pond
Soil Working	rake and hoe	rake and broadfork	rake and broadfork, 2-wheel tractor	2- or 4-wheel tractor, broadfork
Seeding/ Transplanting	hand seeding	Earthway precision seeder	Jang precision seeder	Jang precision or vacuum seeder
Tools	rake, hoe, hori-hori knife	rake, wheel hoe	wheel hoe	transplanter, flame weeder
Soil Fertility	fertilizer, compost, mulch	fertilizer, compost	fertilizer, compost, cover crop	fertilizer, cover crop
Weed Management	hand weeding	hand tool cultivation	hand tool cultivation	mechanical cultivation
Harvest/ Post Harvest	light rinse, not ready-to-eat	field pack to crates, light rinse	ready-to-eat greens, stainless wash tubs, and sinks	compliance with Food Safety Modernization Act and health department
Delivery	personal vehicle	personal vehicle	van	box truck
Markets	direct to consumers (DTC)	direct to consumers (DTC), farmer's market	direct to consumers (DTC), farmer's market, community-supported agriculture (CSA), wholesale	direct to consumers (DTC), farmer's market, community-supported agriculture (CSA), wholesale, retail
Gross Sales	$5k-$10k	$10k-$25k	$25k-$150k	$50k+
Startup Budget	$1k-$3k	$3k-$15k	$20k-$50k	$50k

*Everyone does things a little differently. This chart is based on observations from farms in my area.

the land. This idea is why the mind shift of "starting big" should be reevaluated. If we all wait to farm until we have the means to purchase a large amount of land, we may never start. Our broken food system needs small farms. Our families need the food we intend to grow on those small pieces of property. The notion that you have to have a large plot of land to have a garden is outdated, thank goodness. It should not be about how much land you have. It should be about how much you can positively impact your family and your community in the space you are given.

We have covered small-space values, now let's talk about the budget. We should all have one to some degree. If we do not set a budget for our farm, we will not gauge our profitability correctly or be able to successfully plan expansions without knowing the cost of high tunnels, raised beds, soil, seeds, trays, and other items (see the Farm Cost Analysis chart on the previous page).

Knowing your input cost is vital to the success of your farm. Growing your food may not always seem like the most cost-friendly endeavor, but if you have a budget, it allows you to know exactly where your money is going and how much is left over for other expenses.

I wish I could tell you exactly how to go about this, but I cannot. All farms vary in space, size, and budget. What I may have to spend on seeds, another may not. Someone with the infrastructure already in place will have less cost than someone starting with raw land and building from scratch.

Since I cannot specifically gauge the cost of your farm, I have created these cost analysis breakdowns. Apply the one that matches your farm's needs and size the most.

Carve out time to understand your farm cost analysis, especially if you intend to cut food costs at the grocery store. Doing the math on the front end will help you stay on track with your goals and budget.

We all have the opportunity to do something brilliant when we maximize our small gardening space and work within our budget.

Creating a plan

After you have evaluated your space and budget, you can create a plan for your farm. This can be overwhelming at times, but the rush of excitement outweighs it all for me. When I plan for my garden or farm, I usually write down and dream with a "sky's the limit" mentality. Granted, this does not mean I will have the resources to follow through with everything on my list, but it gives me direction on what I would like to see for the future of my farm. If I intend to build a large workshop on the farm as part of the larger vision, I will make sure not to put permanent bed structures in that area. My husband, Nathan, and I hired a farm consultant for our previous property, and he was worth his weight in gold.

The farm consultant spent an entire day walking with us around our farm, talking through goals, and helping us set realistic expectations for what we could achieve on our farm. If creating a plan is too daunting to do on your own, hire it out. The outside perspective from a professional is usually worth the cost.

I find it beneficial to create plans using timelines. For instance, we do a 1-year plan, a 5-year plan, and a 10-year plan. Doing this lets us become very specific with our priorities and what we are trying to accomplish. If I spent all my time thinking about the 10-year plan, I would likely become discouraged because it is a long-term goal, but if I spend my time working diligently towards the 1-year goal, which eventually gets me to the 10-year goal, I find myself refreshed throughout the process. Plus, let's be honest, who does not love taking a big marker stroke right through a list after you have finally finished it? I'll talk more about hiring a farm consultant and setting 1-year, 5-year, and 10-year plans in chapter 9.

CULTIVATING COMMUNITY

Beets, carrots, and radishes are excellent storage crops. You can grow them in successions, "tip and tail" them when you harvest (cut the tip and tail off the plant), and store them in bags in the fridge for 6–12 months depending on the variety. These are excellent when you are overwhelmed with the summer garden, as little to no preservation is required. Simply pull them out of the fridge when you are ready to use or eat them.

Mentorship is such an excellent opportunity to become invested in your community. When you have experienced hands-on learning through a working mentorship, it connects you. It connects you to the farmer who is mentoring you, you become close to the environment around you, and ultimately, it's connecting you to the things you are trying to accomplish—one day down the road. When I started my farm, I knew mentoring would be something I also wanted to extend to others as I became more experienced. By being mentored by other growers, I saw up close how valuable my relationship was with my community. Be willing to learn and grow within your community, and don't be afraid to eventually be the one who is helping someone else learn and grow.

Grow with your experience

I can express to you all the benefits of being a small-scale grower. I can even help you work through questions to ask yourself about the success of your garden and farm, but I cannot create experiences for you. That is something only you can do, and those experiences will shape how your garden, farm, and you will ultimately succeed.

The best way to learn how to do something is by doing it. Working with a mentor through other farms is always my advice to someone who is just starting. I wish I had done more of this in the beginning years of my journey.

Strive to seek out opportunities to grow your experience and deepen your knowledge. Don't wait for the perfect opportunity to expand your gardening knowledge. You have to seek it out. Continuously search for opportunities to better yourself and your farm. You are only as valuable as the resources you allow yourself access to. When you seek out opportunity, you will learn and develop as a grower. What could take you years to learn on your own, you could learn in half the time through community connections at the core of your focus.

Lastly, do not be afraid to ask. I love giving advice and find great joy when someone comes to me with questions. First, I love that they want to learn, and secondly, I love that they are unafraid of rejection and brave enough to have asked in the first place. To be candid, you will be told *no*; I was told *no* more times in the beginning than I care to remember. Still, I continued to ask farmers if I could volunteer. I continued to ask any gardener willing to give me the time of day for advice. Eventually, I found the right people who believed in my vision and were ready to do whatever was necessary to help me get to that vision. I discovered that plenty of people were ready to say yes to my requests and questions—despite the people who told me *no*.

Now that we have talked about why small-farm values are so important to me, and hopefully you as well, my hope is that the rest of this book equips you to make your tiny farm efficient, productive, and fruitful.

Tiny tip

Remember, good things take time. There is much value in dreaming and working hard towards your goal, but remember to stay the course and keep focused on what is directly in front of you. One of the biggest hurdles with gardening is simply getting started. Set realistic expectations for yourself—there is no such thing as a perfect garden. Plants are going to die, weeds are going to grow, you may have to redo your hard work at times and troubleshoot different challenges that present themselves. Don't despair, just keep growing.

I want to challenge you to start growing your community today. Be honest and create real relationships that you can lean upon as you grow and continue to pursue your vision.

I'll talk to you in chapter 2 about the types of farmers, and you'll discover more about what kind of small farmer you would like to become.

When we extend an invitation for our children to explore our passions, that passion overflows to them.

CHAPTER TWO

What Kind of Farmer and Gardener Do You Want To Be?

I want you to imagine your ideal garden. Now imagine yourself in that garden. If it helps, close your eyes to visualize the result. What do you see? A hobbyist? An aspiring farmer? A pillar of sustainability for your family? Whatever picture pops up into your mind, jot the description down on a piece of paper, and keep it. These different types of farms all have a purpose and space; the question is, which one are you? The things you wrote down will help define you as a gardener and farmer.

—————————————— ☼ ——————————————

When I first began gardening, I did this visualization practice and vividly remember seeing my children running up and down the garden paths. I recall harvesting buckets of produce and stocking the pantry. I imagined myself being so desperate for this sustainable lifestyle that I would do just about anything to get there. The "why" in the beginning was to learn more and be a good student. I knew if I were teachable, I would see the produce buckets start rolling in and the pantry filling up.

Over time, though, my "why" changed. Now, my reason is to provide food for my family and to teach my community. Now, I grow high food yields on a small plot of land and my goal is to teach others how to produce an abundance of food themselves without having hundreds of acres.

We naturally express our values when we are rooted in our belief systems. They anchor us to truth and help propel us forward.

You will never understand the enjoyment of growing food until you do it. Plant the seed and watch it grow.

I will make reference to remembering the "why" throughout this book because it's essential. When we forget why we wanted to become farmers and gardeners in the beginning, we find ourselves in a season of burnout.

Expressing values

I hope you have an idea of what you envision for yourself as a grower. Now, ask yourself how you want your values to be expressed? Will you be expressing your values well enough through your lifestyle and how you practice sustainability on your farm so that others can adopt the same principles and apply them? What values do you hope will come to mind when someone thinks of you?

My hope for you is that through this book, you will understand the importance of:

1. Investing in yourself and your knowledge

2. Knowing we are all a piece of a much larger puzzle. Understanding that each of us, despite our scale, is connected. Unified, we all can make a tremendous impact on the Earth that surrounds us.

Let's steward our Earth well, friends.

Knowing your influence

It is important to evaluate the influence you want to have as someone who grows food. When I talk about influence, I'm not referring to how many people are following your journey. I am talking about making an impact and knowing why you chose to pursue this. It's vital to assess the type of impact and influence you want to have.

You can influence your family by getting them involved in the garden. Raise your children to know where their food comes from and how it is grown.

You can influence your friends and extended family by sharing the bounty of your harvests and expressing how purposeful this life is.

You can also influence people you've never met simply by living out your life authentically and sharing the ups and downs of your journey with them.

Learn to grow with your community. Relatability is a powerful tool. Sometimes our most significant influence is the ability to listen to others. My sweet daughter, Charlee, teaches me so many things when I am willing to pause and listen to what she has to say.

Take moments to stop and pause. Reflect on your journey to becoming a gardener. Think of how far you've come and the impact you have on those around you. Who knew many moons ago my love of growing food would lead me to having these tiny dirty feet in my garden? What a beautiful journey it has been.

Methods of approach

The methods you choose to implement on your farm will define the kind of gardener or farmer you want to be. No pressure, but this is something you will want to think about long and hard before jumping in.

Is sustainability a driving factor behind your reason for farming or gardening? Do you care about putting practices in place that will promote the longevity and health of our Earth?

Are you looking for a quick fix or determined to help fix a much more significant problem? There are no easy, overnight methods that will yield impactful results. But there are ways that serve a greater purpose, and the time involved in building those is worth every hour of effort. If you want to be a part of the solution, it will take time.

Continuously improving

You will continue to evolve as a grower through the years, and your systems and influence will too. Be mindful that there is no right or wrong way to farm; we all run our farms based on our convictions, and I believe that is how it should be. If we all modeled the same, life would be boring. We would all lack our identities and unique values. Dare to be different and seek out what you feel is best. This book is merely a tool to teach you the styles and resources of farming, but you have to use the opportunity to make your tiny farm your own; embrace that.

Did you know that each sunflower is thousands of tiny flowers? The petals and brown centers are all individual flowers themselves.

Gaining perspective

You won't know what kind of farmer or gardener you'll want to be right away. It's going to take time, and more than likely, the type will change along the way. I am certainly not the same gardener as I was when I started, and for that, I am thankful.

I did not know a single thing about gardening, farming, seeds, or soil when I started. But over time, I began to realize my goals and the steps I could take to achieve those goals, and I began to move forward, no longer continuously taking backward steps.

If there is one thing that will stick with you through this entire book, let this be it. **Set realistic expectations for yourself**. If you can figure out how to set your expectations so you do not add more than you can handle, you will have found the key to success.

> If you can figure out how to set your expectations so you do not add more than you can handle, you will have found the key to success.

Sometimes it's best to take a step back and gain perspective. Have you accomplished what you set out to do? Did you grow the amount of food you had hoped to? Are you taking steps forward or moving backward?

When you start evaluating what you are doing, you might realize that your whys have changed, and therefore, the direction of your farm needs to change. Some people may desire to go back to a slower pace, while others may feel the yearning to plow ahead, maximizing space and production.

The importance of the garden

Gardens are important for several reasons; food sustainability, access to healthier fruits and vegetables, and let's not forget the positive impact (when cultivated correctly) that it has on our planet.

More than likely, most people you encounter would agree that there are more pros than cons when it comes to growing your own food. But, the real question here would be, what is the importance of the garden to *you*?

I'll go first. On our farm, we have what we call "Farm Pillars," which remind us—and anchor us—to what and why we do what we do. Number one on the list and the center of everything we do is "Community." Don't worry, growing food is second! For us, the importance of gardening is more about the impact it will have on the community, whether it be through education, produce, or inspiration, rather than through the actual food itself.

You don't have to be an expert to grow food. You don't have to be an expert to make an impact. You simply need to have a willing spirit and listen to what the garden has to teach you.

Radishes are the perfect crop to grow for beginner or seasoned growers. These purple daikon radishes will last for several months stored in a refrigerator. Add a beautiful pop of color in kimchi, or they're perfect for fresh eating.

When we grow our food organically, it reduces pollution, reduces soil erosion, increases the fertility in your soil, and is also perfect for afternoon snacks without the thought of harsh chemicals being used.

Avoiding burnout

Avoiding burnout is crucial to your success as a grower. It's essential to create spaces throughout your farm that help you rest, but also keeping realistic expectations in mind is vital. Be like an anchor, steady, unwavering, and holding onto what matters.

Be like an anchor, steady, unwavering & holding onto what matters.

Now I wouldn't be writing a book about growing food if I didn't believe in the value of growing food, but this is the message I am trying to get across.

I can grow food and keep that food for myself and my family, or I can grow food and teach, educate, and inspire others to grow food for their families, and you can already see the ripple effect that starts to take place.

I want to use my influence well, and I want this book to equip you with all the knowledge you need to grow abundant food. Remember the influence we talked about earlier in this chapter? Here's a good time to reference that.

What kind of impact can your tiny farm have?

Don't try to keep up with the Joneses. You have to know *your* reason why you farm and sticking to *your* values is a key factor. There is no one way to do this. Start small. Take baby steps each day that will get you to your long-term goal. You have to create space to enjoy the fruits of your labor. I put things in the garden just for fun, and I make time to go to the garden even when I'm not working. The cottage garden and raised beds are places I simply go to rest. Even when I do feel overwhelmed, I recall the pillars plastered on my dining room walls. This is what I'm called to do, and this is what pushes me through seasons of burnout.

Annuals, biennials, and perennials

What is an annual?

1. Annual plants experience their entire lifecycle, from germination through flowering and fruiting, setting seed, and then dying, all within a single growing season.

2. They are replanted each growing season.

3. Most vegetables that you see at the grocery store or farmer's markets fall into the annual category. While there are plenty of vegetables that are true perennials, we still grow them as annuals because they can't withstand the cold temperatures of winter. So, unless you are growing them in a heated greenhouse, throughout the winter, most perennial vegetables, like tomatoes and peppers, will be grown as an annual.

What is a biennial?

1. Biennials are plants that have a 2-year lifecycle. In their first year, they grow leaves, stems, and roots. Then they go dormant for the winter. In the second year, they flower and produce seed.

2. Biennials are common in flowers, but you'll also find vegetables on this list as well. Biennial vegetables are typically grown as annuals because we want to harvest them before they flower and set seed (unless we are growing them for seed saving, of course).

3. Some vegetables that are biennial include: beets, brussels sprouts, cabbage, carrots, celery, kale, rutabaga, swiss chard, and turnips.

What is a perennial?

1. Perennials are plants whose lifecycle lasts for more than 2 years.

2. They can be planted from seed, bulb, or plant divisions.

3. Perennials are less work than annuals. They go dormant through the winter instead of dying and redirect their energy into establishing healthy, strong roots for the following spring season.

Why grow edible perennials?

Emphasize perennials for your farm. I have this mission. It may seem silly to some, but it will be life-giving to others. Suppose we, as growers, start making a place for edible perennial crops. If we do, we will see our soil quality improve because these plants stay in the ground longer than one season, we aren't leaving our soil exposed to soil erosion issues and compaction. We reduce our need for more inputs. And then the best part—they will produce delicious yields year after year.

I am aware that if you are pursuing a vegetable garden, you may be limited in the number of perennials you can include, so here's what I suggest: Add whatever perennials you can. For us, that is asparagus, various blackberries, and raspberries. We also include perennial herbs to use fresh in the kitchen, and we can't forget the perennials that we grow for cut flowers.

Perennials are an extremely sustainable way to grow, and I think if we can all emphasize the importance they have on a farm, more growers will see their benefits too.

Perennials for the win

Create practical spaces throughout your farm that serve more than one purpose. In the raised bed garden and high tunnels I don't have the space needed to grow many berries or perennials, but my cottage garden is the perfect eclectic space for these additions to our farm. It is a space that I can let go a little wild without feeling the guilt of lack of production.

You might have read the reference to my cottage garden above and thought to yourself, what is a cottage garden? Well, a **cottage garden** is a style of garden that implements informal design, a mixture of ornamentals and edibles, and is composed of traditional materials.

I've asked several people what they think of when someone says cottage garden, and almost every time, the answer I receive is, "Isn't that a garden in England?"

While my cottage garden isn't quite like the description above, it is a place that's eclectic and peaceful for me to go to find rest. My cottage garden has a quaint glass greenhouse and a mix of berries, edibles, and perennials.

When planning out our cottage garden, my goal was to establish as many perennials as possible.

Nurture yourself. Speak kindness and pour back into yourself like you would others.

Push your limits

We all have those moments where our fears and insecurities limit us. It truly happens even to the best of the best. No one is off-limits. The difference is, though, which of us will choose to push past those fears and jump in, and which of us will be crippled by them? You don't have to be an expert to grow food: you just have to be willing to push past your limits even when it's uncomfortable.

You don't have to be an expert to grow food: you just have to be willing to push past your limits even when it's uncomfortable.

I prefer to live with a legacy mindset, and to do so, it requires something of me. So, I say take the leap of faith on a new piece of property or start growing food for the first time. Push your limits even when you're scared. Plant the seed, and pray it sprouts.

I tend to find a lot of growth in that in-between place of being scared to death and knowing success is right around the corner. Lean into that feeling, and I can promise it almost always pays off.

Starting seeds will push you out of your comfort zone and grow you as a gardener. I recommend starting seeds you are familiar with and learning as you go.

Create a sanctuary

I've learned and grown so much as a gardener through the years, from tilling my backyard and planting my first garden to cultivating raised beds at my second home, growing in high tunnels, starting no-till beds, and maintaining a massive, raised bed garden. Though the style of the garden and how I chose to grow in it has changed quite a bit, one thing always remained.

I created a *sanctuary*—a place where I find solitude. I can walk out there any time of day and listen to the birds sing a glorious song. Or hear the frogs croaking at night. I laugh in my garden and shed tears over ruined crops. But most importantly, the garden is a place of refuge. I see creation all around in full bloom, thriving plants, and bountiful fruits—it has become a sanctuary I can't live without.

What is a sanctuary?

A sanctuary is a safe place or a place of refuge. I am trying to create a place I can reside in after a long day's work. This is a place I can retreat to when I need inspiration or to spark creativity. It is a space I gravitate to simply to rest. In my opinion, every farm needs a space like this. Trust me. You'll thank me later.

Your sanctuary will look different than mine. It may be planting tomatoes in May or harvesting sunflowers in the summer heat. Maybe it's enjoying a fresh glass of lemonade in a space you've poured so much of your heart into regardless of why you choose to come back to the garden.

We should create a garden space that draw us to come back. Creating such a space will remind you daily of your journey, this process, and to never give up on your dreams of growing food. Every victory, no matter how big or small, is worth celebrating.

As simple as it may seem, paint the garden gate, hang up the welcome sign, create a space you feel drawn to come to day in and day out.

The good life

When we think of farmers, we usually think of how overworked and underpaid they are, especially the small-scale grower. It is no easy feat to create a business out of growing food, but I am here to tell you it doesn't always have to be about working seven days a week nonstop (although, if I were honest, this is still something I'm figuring out).

If you are venturing out on your own food-growing journey for the first time or you're a seasoned grower, embrace the good life.

Yes, it's hard work. Yes, it's continuous work. But my goodness, it is one of the most valuable things we have the privilege to do. I encourage you to remember that, embrace that, and practice, what is, "the good life."

It's about so much more than growing food. Farming is a lifestyle we choose to live every single day. Farmers are very much married to our farms, and that's how we like it. This life we've created is sweet, meaningful, and full of purpose.

Please don't hear me wrong. We still have friends, my kids still have activities, and I will always prioritize date nights with my husband. But our farm is our baby, our livelihood, and the drive behind our passion and our reason why.

It doesn't have to be like this for you. Farming isn't this "all or you fail" type of thing. But for some of us, it will be our careers and require more work and attention, and for others, it will be a fun hobby that takes a few hours out of the week.

The relationships you pour yourself into are one thing you'll seed that continues to grow well after the frost has set in.

Spinach is a must-grow winter staple. It can withstand temperatures as low as 15° Fahrenheit (-9° C). It is packed full of nutrients, and can be consumed raw, cooked, and even preserved.

Cultivate what matters

I sat down with my papaw this week, and we talked about the good ol' days—the days where I worked with him nonstop in his garden. I remember setting up the pressure canners outside over a propane burner and making jelly for what seemed like days.

I was around seven years old when I was working with my papaw. He would "hire" us to weed the garden (worst job ever) and help him harvest tomatoes, blueberries, grapes, purple hull peas—you name it! My brother and I would pretend that weeds were people we didn't like, so that was our motivation for getting our job done. After all that hard labor was done, and we'd receive our couple of bucks for hours worked, I'd run inside to my memaw Jean, throw on my favorite apron, and we would cook together. Those were the days when I wanted to be a chef. Haven't you heard of the *Cooking with Jill* show? Well, that's a shame. I was the star of that show for an audience of three people.

I had no idea that those weeds, sore hands, and steamy summer days were shaping my future as a farmer. I look back on those moments of what I

thought were useless, boring lessons, and I'm filled with gratitude that my mentoring started at such a young age.

My papaw tells me stories from when he was a kid and how you grew a garden more out of necessity than desire. He said it was simple: if you didn't grow your food, you didn't eat.

He recently sent me home with black pole beans that he swore were the best I'd ever eat, heirloom tomato seeds, and a heart full of memories.

At 83 years of age and full of wisdom, my papaw is still teaching me to cultivate what matters most—relationships.

Each season, your garden will fade. There will be a time when you have to reseed and regrow, again and again. The relationships you pour yourself into are one thing you'll seed that continues to grow well after the frost has set in.

If your why is to farm for your family, cultivate those relationships well, and encourage them to become involved. If the community is your driving factor, pour back into them and invite them into your gardens and your homes.

The greatest thing you'll cultivate on your farm is what matters to you.

CULTIVATING SELF

irlooms will always have a special
ace in my heart and on my farm
cause of the sweet memories passed
wn by my papaw years and years
o. This 'Kellogg Breakfast' is
e of the fruitiest, flavor-packed
matoes I've ever eaten.

Investing in yourself looks different for many of us. For me, it means spending time in the garden when there isn't work to be done (or should I say when I have no intentions of working). I have my cup of coffee under the pavilion in the raised bed garden every morning. It's my quiet place and my sanctuary—it is where I am surrounded by the beauty and awe of what I've been given to steward. I'm left feeling grateful and rejuvenated by the time I finish off my cup. Listening to the birds chirping and seeing the sun peeking over the trees as it shines down over freshly planted brassicas fills me up in a way I could never honestly describe. It keeps me moving forward. It's how I serve myself so that I can pour into the garden.

Nurture the gardener

Often the one who gets forgotten about in the garden is the gardener. I've learned over time that if I don't tend to myself (the gardener), there will be no garden. You, sweet friend, are valuable—the most valuable of all.

In this chapter there are some tough questions that will require you to think hard about your goals and what kind of farmer or gardener you want to be. Don't lose sight of why you started, find the rainbow in the storm, and keep moving forward.

Types of growers

Knowing what kind of grower you are will help set you up for success. You will be able to easily establish rhythms and routines and set realistic goals for yourself as you journey into this gardening adventure. Read through each description below and see which one resonates with you the most. Disclaimer: there will not be a perfect match. Instead, focus on which lifestyle you most relate to. This assignment is not to put you in a box, but rather to help guide you so you can set realistic goals and grow with purpose.

You will find a few things sprinkled throughout my garden that bring me joy. These giant marigold flowers do just that. They are one of the most fragrant flowers I've ever smelled. They look like sunshine, and make my heart so happy every time I walk by.

The hobbyist

This was me 8 years ago. Working a full-time job, growing food in a few 4 x 4-foot (1.2 x 1.2 m) beds in my backyard, but primarily still buying the bulk of our family's food needs from the grocery store and farmer's market. I gardened because I enjoyed getting my hands dirty. Food interested me. I was in a phase of my life where I wanted everything to be handmade from scratch and grown as environmentally friendly as possible (turns out, this phase wasn't just a phase but how I structure my life to this day.)

I felt accomplished every time I would can another jar of pickles or put up jelly from peaches we got at the orchard (discounted because they'd fallen on the ground and weren't okay to sell to the consumer anymore).

We worked with a small budget to add more beds here and there, but the garden stayed small, and my time to tend it was even smaller. I only grew food during the summer, when I had more time in the afternoons after work to check on things, but the weekends were when we would put in the work (harvesting, canning, etc.).

Brassicas like kale and cabbages are easy hands-off crops. Once planted, they require little continual maintenance, aside from the occasional hand-picking of caterpillars.

Ultimately, gardening as a hobby was a fun and peaceful place. I would find though, that the hobby would eventually grow into something much more.

The hobbyist

- Buys from the farmer's market

- Has a few raised beds in the backyard

- Works a 9 to 5 job

- Finds great joy in gardening

- Works on the garden in the afternoons and weekends

- Finds that gardening helps you unwind, connects you with nature, and brings you peace

The homesteader

The homesteader (or aspiring homesteader) is that sweet spot between hobbyist and farmer. When I think of a homestead, my mind immediately jolts to someone who is self-sufficient. Your goals aren't to grow food for market and profit. Your aim is to become sustainable—growing food solely for your family's needs.

You may work outside the homestead, or perhaps the homestead is your livelihood. Along with growing your produce, you raise your own animal protein (or if you are a plant-based homestead, you grow more produce) and have dairy animals on your farm.

Most of your cooking is from scratch, with all the ingredients grown or raised on the farm. Your quest is to be self-reliant and sustainable.

I know many homesteaders, and it is an all-day, every day show-up and work responsibility. But it's also a passion, and with that passion comes the drive to get up and pour back into your homestead.

The homesteader

- Has a goal that is not profit—it's sustainability
- Might have another job or might homestead full time
- Raises most of your protein and vegetables on the farm
- Cooks from scratch
- Strives for sustainability

'Grand Marshal' is a hybrid variety intended for greenhouse production. Its fruit is uniform, disease resistant, and withstands the humidity of a high tunnel with ease.

Typically, these are market-style growers, maxing out production. The main goal for the farmer is profitability. Without profit, they have no job.

The farmer

- Farms full time
- Makes a living off their farm
- Aims to be profitable
- Runs a community-supported agriculture (CSA)
- Sets up at the farmer's market
- Profitability is your focus

Learning to eat with the seasons makes a significant impact on your ability to grow and consume more food from your farm. Summertime is meant for fresh tomatoes, abundant cucumbers, and hand-picked green beans.

The farmer

Lastly, we have the farmer (or aspiring farmer, perhaps). I spent years working towards this, and I wanted nothing more than to be farming full time as my career. It looks a bit different for me than some of the things listed below, but it is what I am doing full time, and it's a passion I'll forever carry with me.

For most, though, the farmer is pursuing this life full time, and it is their primary income. They set up at farmer's markets every week to sell their produce or provide a CSA subscription box.

What is a CSA?

CSA stands for community-supported agriculture, and essentially it is a subscription box customers buy into. In exchange for their membership, they receive either weekly or bi-weekly shares of a box full of fruits or vegetables that were grown by the farmer.

Now that you know what kind of grower you are, it's time to create your personal plan so that you can grow with purpose.

'Katrina' cucumbers are pretty special. They are *parthencarpic*, which is fancy for meaning they don't need to be pollinated. Pretty cool, right?

CHAPTER THREE

Grow With Purpose

All right, friends, buckle up. We are about to get into the nitty-gritty. We've discussed why small farms are indispensable to the health of our local environments and our communities as a whole. We have evaluated what kind of farmer and gardener you are (or want to be). Now, we are about to dive into market gardening, growing your food with purpose, and some things to evaluate and implement to ensure that you make the most out of your tiny but mighty farm. So, let's dive in.

Throughout this chapter, we are going to discuss the market garden.

What is a market garden?

A market garden is a small plot of land, usually from one to a couple of acres, where fruits, vegetables, and flowers are grown and sold to the consumer. You often hear what's grown in a market garden being referred to as a cash crop. This means the crops are being grown with the intent to sell them to the public through farmer's markets, restaurants, and CSA shares (more on CSAs in chapter 9).

Most market gardens include greenhouses or other types of covered growing space where the growers are continuously starting and seeding plants to provide their consumers with products through the growing season.

Unlike large-scale farms that rely on monoculture, market garden farmers grow a variety of crops, most of which are grown using hand labor and usually following organic gardening practices.

The key to productive and successful market gardening is efficiency. The goal of being efficient means eliminating wasted steps, wasted time, and ultimately wasted resources. In chapter 7, we will cover efficient systems and we'll talk more in-depth about practical ways to streamline your work and productivity.

You don't have to sell your produce to be considered a market gardener. It is more about the practices you put in place to be efficient and grow food abundantly.

Coupled with efficiency is effectiveness. It's a powerful word that means just as much to a grower as being productive. Sure, the goal is to put in place adequate systems that save you time, but we also have to become effective with our approach and implement it daily. For me, that means making a list at the beginning of each day, prioritizing what needs to be done and in what order.

If the big chore for the day is cultivating all the raised beds, you won't find me doing this by hand. Instead, I will use the proper tool or technique so I get the job done in less time. Getting an outside set of eyes to examine your process is always helpful in assessing whether there is an opportunity to save time by using a better tool or technique.

When you create lists for yourself, the expectation is set. You know what you need to get done that day and can map out your time accordingly.

Choosing the proper tools to get the job done saves you time, energy, and ultimately money in the long run. Think smarter not harder when it comes to being efficient on your farm.

Along with seeders are dibbler plates. These are great for making indentions for your seeds to easily drop into. This makes a long day seeding in the greenhouse go by much faster.

Let's say it's a seed-starting day in the greenhouse and I have thousands of seeds to sow. If I were hand seeding one by one, this job alone could take me all day and then some to accomplish it, but if I invest in the proper tool and use a drop seeder, I am able to seed an entire 50-, 72-, or 128-cell seeding tray in no time at all.

What is a drop seeder?

A manual drop seeder is simply a frame with interchangeable plates. Depending on the tray size you have to seed, the drop seeder can be adjusted to various sizes and widths and allow you to seed an entire tray at once.

By growing with purpose and efficiency at the forefront of our minds, we can spend more time enjoying the fruits of our labor as a family.

WHAT IS WHAT

You may have heard the phrases market gardening, micro-farming, and hobby farming, but what do all of these terms mean?

Market gardening: Intensive production on a small plot of land, usually less than a couple of acres, where fruits, vegetables, and flowers are grown and sold to the consumer.

Micro-farming: Small-scale, high-yield, sustainably minded farming, generally conducted by hand in an urban setting. Micro-farming can be for profit or for pleasure.

Hobby farming: Small-scale farming that is conducted for pleasure and not as a source of income.

A place for technology

In our 3-foot (1-meter) beds, we place irrigation lines every 12 inches (30 cm).

We are able to spend more time enjoying our farm and less time managing it by using technology to help automate various chores. For example, by placing drip irrigation in all of our raised beds, we have not only taken the guesswork out of watering, but we have also saved a significant amount of time walking back and forth to turn the water off and on.

Furthermore, we can leave the farm for the weekend and know things are cared for. In preparation for our departure, the drip irrigation systems are hooked to a timer set to go off in the morning and the afternoon.

We also place fertilizer injectors on the frost-free yard hydrants that connect to our irrigation system. These are set on a timer to go off monthly, feeding our plants and improving our soil's ecosystem. By doing this, I no longer have to spend hours (or even days) adding prebiotic and organic fertilizers to our soil.

You can change or implement new ideas on your farm continuously. Keep trying new things until you've figured out what works best for you and your schedule.

Establishing land (and a plan)

Now that we've detailed what a market garden is, let's further discuss the importance of finding suitable land.

Something about being one with the land speaks to many of us—desiring a quaint spot in the country. Maybe we daydream about the perfect farm with a little house, animals in the back, and gardens as far as the eye can see. Or perhaps it's memories from childhood of our grandparents' farm.

The memory of my papaw's farm will forever be ingrained in my mind. It certainly wasn't what most farms look like today, but there was more space to run free than you could imagine. I would walk out his back stairs, and to my right was the largest garden. Before you reached the garden, though, you'd have to pass through long rows of trellised up grapes, and then behind that was, of course, all the other berries he grew, like, blackberries and blueberries. Then, you'd finally reach the in-ground garden with row after row of all the summer vegetables. He had a big shade tree not too far from there, with a picnic table and a swing. It wasn't just a garden. He created an atmosphere that I carry with me to this day.

Dreaming of this is the easy part. Finding this perfect slice of heaven can be much more challenging. I would suggest not looking to obtain the perfect spot but instead creating it yourself.

> ## I would suggest not looking to obtain the perfect spot but instead creating it yourself.

You may already be living on your perfect piece of land or you're in the stages of trying to buy it. Think of your property either now or in the future as a canvas. On it you have the opportunity to paint a detailed picture unique to your needs and desires.

When searching for your piece of paradise, something to consider is what you envision your farm and gardens to look like. Are you hoping to achieve market-style rows as far as your eye can see? If so, expanses of pastures will be high on your priority list. If you are hoping to use animals to prep the soil for you, then look to see if fences are on the property or will you need to account for the extra cost to install them? Having a clear vision for your space before purchasing it is extremely beneficial in moving forward with the decision-making process.

When you have a clear picture of what you envision for your space, you will now be able to immediately begin working towards your goals.

Bigger doesn't equal better

Many years ago, small-scale farmers did not exist. Thankfully that is not the case today. Pioneers like Eliot Coleman taught us the value of growing on a small scale and just how purposefully it can be done.

Small farms have all but disappeared over the past 100 years, mainly due to industrialization and consolidation. Now, a small number of mega-farms and corporations produce and process the vast majority of our food. Luckily some leaders in small ag have raised the warning flag on the effects of that trend and have led the way in developing and sharing techniques to help small farms prosper

and put food production back in the hands of the local communities. In addition to Eliot Coleman, pioneers and trailblazers like Jean-Martin Fortier and Ben Hartman have taught us the value of growing small and just how big an impact it can have.

The idea of buying more land, investing in more infrastructure, and building additional raised beds is not the only way to grow higher food yields. Yes, we live on a few acres, but the amount designated strictly for food production is less than an acre, and if I were really maxing out yields in a traditional market garden, I would need even less space than that. Here's how we do it, realizing that less is more. Less infrastructure means less upfront cost on your farm, and if you are on a tight budget, this is probably speaking to you right about now. If we figure out our plan for the space and the budget we have (which we will further discuss later in this chapter), we can start to implement things like succession sowing, which really gives us the freedom to do more with less.

Choosing the right varieties makes a world of difference in how productive your tiny farm is. These 'Sakura' cherry tomatoes produce abundant fruit and have a meaty taste with a sweet note at the end.

What is succession sowing?

Be mindful when starting in soil blocks this small to water regularly. Because they are so small they dry out quicker. In the summertime I am making frequent stops in the greenhouse and watering more often.

Succession sowing is a way to extend your harvest throughout the season by staggering your plantings. You do this by seeding your crops at intervals that range from 7 to 21 days, depending on what you are seeding. Once you harvest your first crop, you immediately plant your second one and continue this process.

Two things we have started to succession sow throughout the season are baby greens and spinach. These are the two main crops we focus on because they are the crops we are eating year-round, meaning if I am not continuously growing it, then I am spending money on getting it elsewhere.

Here is the chart we use to help us time our succession plantings. You can modify it for your farm, just list your crops, and add your planting or sowing dates and average first fall frost date.

Succession Planting Spreadsheet

Fall Frost Date:											
Variety	Days to Maturity	Days Between Succession Plantings	1st Planting	2nd Planting	3rd Planting	4th Planting	5th Planting	6th Planting	7th Planting	8th Planting	Final Planting Date
Basil	65-70	14									
Beans	55	10									
Beets	50	14									
Cucumbers	60	21									
Lettuce, Head	55	14									
Lettuce, Baby Leaf	30-35	7-14									
Radishes	26	7									
Spinach	40	7-14									
Summer Squash	50-60	42									
Sunflowers	45-50	7									
Zinnias	60-70	14									

Interplanting is a great way to maximize a small space. You can interplant root vegetables like radish, beets, and carrots. Lettuce and kale are great for this, too.

Another way we can make the most of our small space is by putting a large emphasis on our crop plans and evaluating the appropriate place and space we have for each crop. If you intend to have a couple of raised beds in your backyard or have most of your property designated for food production, a crop plan will be extremely beneficial.

One, it gives you a visual idea of what your space will look like. For me, this is important. I can erase and rewrite as many times as I want until I've landed on the perfect plan. Without a crop plan, once you've planted a bed, there is no going back.

And, secondly, a crop plan helps keep you organized and on a schedule. They can be as detailed or as simple as you'd like. We have an additional spreadsheet that tells us when we are starting and transplanting our crops, but with this general crop plan, you can be prepared for the amount and type of seeds you'll need to buy for the season, know when to start and transplant them, and it can be a valuable record to have on hand in the upcoming years as a reference. Once

you've put these practices in place, they will become second nature and give you the freedom to focus on other aspects of your farm.

Develop a rotation plan

Draw a rough map of your farm's planting beds and use it as a template to determine your crop rotation plan for each season. These four plans represent what will be planted in each of our beds during the four seasons of a single year. Your rotation plan will look different from ours because it is based on your climate, frost dates, the layout of your garden, and the crops you want to grow. And every year, your plan can be different as you rotate crops to new beds and grow different things.

Winter rotation plan

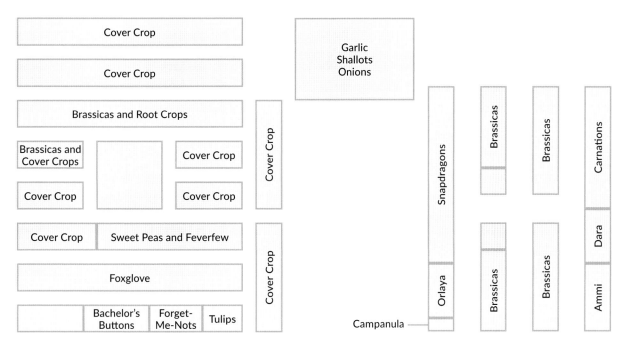

Cover Crop

Cover Crop

Brassicas and Root Crops

Brassicas and Cover Crops		Cover Crop
Cover Crop		Cover Crop

Cover Crop	Sweet Peas and Feverfew

Foxglove

	Bachelor's Buttons	Forget-Me-Nots	Tulips

Cover Crop

Cover Crop

Garlic Shallots Onions

Snapdragons
Orlaya
Campanula

Brassicas
Brassicas

Brassicas
Brassicas

Carnations
Dara
Ammi

Spring rotation plan

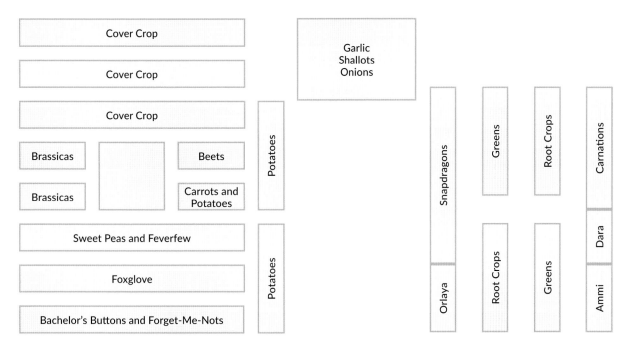

Cover Crop

Cover Crop

Cover Crop

Brassicas		Beets
Brassicas		Carrots and Potatoes

Sweet Peas and Feverfew

Foxglove

Bachelor's Buttons and Forget-Me-Nots

Potatoes

Potatoes

Garlic Shallots Onions

Snapdragons
Orlaya

Greens
Root Crops

Root Crops
Greens

Carnations
Dara
Ammi

Summer rotation plan

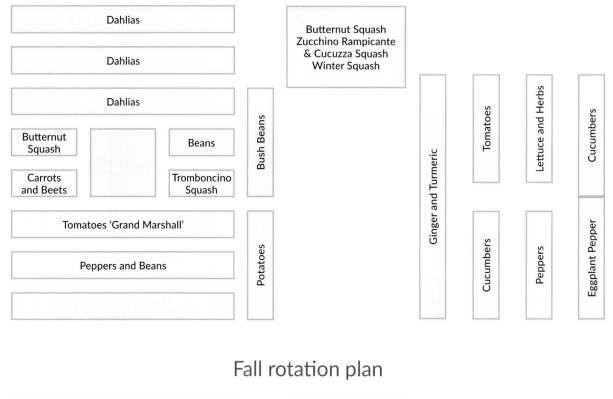

- Dahlias
- Dahlias
- Dahlias
- Butternut Squash
- Carrots and Beets
- Beans
- Tromboncino Squash
- Tomatoes 'Grand Marshall'
- Peppers and Beans
- Bush Beans
- Potatoes
- Butternut Squash, Zucchino Rampicante & Cucuzza Squash, Winter Squash
- Ginger and Turmeric
- Tomatoes
- Cucumbers
- Lettuce and Herbs
- Peppers
- Cucumbers
- Eggplant Pepper

Fall rotation plan

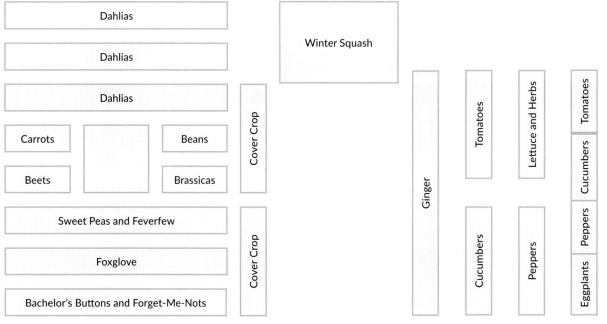

- Dahlias
- Dahlias
- Dahlias
- Carrots
- Beets
- Beans
- Brassicas
- Sweet Peas and Feverfew
- Foxglove
- Bachelor's Buttons and Forget-Me-Nots
- Cover Crop
- Cover Crop
- Winter Squash
- Ginger
- Tomatoes
- Cucumbers
- Lettuce and Herbs
- Peppers
- Eggplants
- Peppers
- Cucumbers
- Tomatoes

Find new ways to make farm chores enjoyable for yourself and your kiddos, if you are trying to get them involved. It never fails that every time the kids help me with watering seedlings, we all end up soaking wet. These are memories we will all treasure for a lifetime.

Tiny tip

Have your crop plan written out before buying your seeds for the season. By doing this you will avoid overspending and buying unnecessary seeds you don't plan on using. This is a budget-friendly way that I stay on track.

Grow with purpose

Growing your family's food does not have to be a daunting task. Instead, it can be pretty enjoyable. I was reading a garden article some time back, and I came across a sentence that stuck out to me. It read, "Calling ourselves gardeners implies we have good intentions for the garden, when in fact, it means nothing about intentions, but more about purpose."

At first, I raised an eyebrow. What do you mean calling ourselves gardeners means we are not being intentional? The garden is built on intentions. An entire garden is an act of intentionally planting, weeding, and harvesting.

> "Calling ourselves gardeners implies we have good intentions for the garden, when in fact, it means nothing about intentions, but more about purpose."

I wrestled with this concept for some time, and then, one day, it hit me.

It was a hot Arkansas July day, and I was harvesting tomatoes. I somehow convinced my sweet daughter to help me. I'm not sure if it was pity or being enticed with a few bucks that got her to agree, but whatever the reason, I was thankful.

I've grown a lot of things in my lifetime, but these girls are some of the sweetest.

An hour into harvesting, sweat dripped from our beaming red faces, and we were worn out and tired of it. Charlee looked at me and said, "*Mom, do we do this because it's meaningful?*"

Cue the lightbulb. My daughter, young and brilliant, understood the purpose behind growing our food. Sure, she might have questioned it at the moment. Heck, I even questioned why I chose the hottest day of the year for tomato harvesting. But, at that moment, I realized it was about much more than growing with intention. It was feeling purposeful about the work you are doing.

Any grower of any scale can intentionally grow food, but not everyone feels purpose in why they are doing it, and now I realize *that* is actually what makes a garden thrive.

This was a beautiful opportunity for me to pour into my sweet Charlee and explain my "why" behind gardening.

Remember a few chapters back when I mentioned how I would reference your why throughout this book? Well, here it is, friends. Everyone's why is different, everyone's purpose for growing food is different, but here are some questions to ask yourself when learning how to grow food with purpose.

1. Do you enjoy growing food for yourself and your family?

2. What is the driving factor that motivates you?

3. Is your plan sustainable?

Once you've answered these questions, you can fall back on them and remember why you are waking up every day, putting one foot in front of the other, and continuing on your food-growing journey. When I experience burnout, I remind myself why I enjoy growing food for my family, I reference my community—which is a huge motivator for my why—and I remind myself that this is sustainable and purposeful. I encourage you all to do the same.

Creating a strategy (for your family)

I grew our food primarily in raised beds and would sell particular crops to consumers. While we would take produce to local restaurants, we did not primarily grow our food for that purpose. And now we do not sell any of our produce to consumers at all. We aim to put up a year's worth of produce for our family and donate any excess to the local food banks. However, we still want to utilize the things most market farmers implement for production purposes. Our main goal is to produce a large amount of food as efficiently as possible. And to show our community how we grow and manage our farm.

When growing your family's food on a farm and with the intention to potentially sell the excess, the first thing I recommend is to craft a strategy. (I know, I know not all of us are gifted at being organized, and I'm talking to myself here, but I learned that it was necessary for production on our farm.)

I encourage you to sit down with your notepad and write out your goals for the season. Is your goal to grow the ingredients so you can preserve enough salsa to last an entire year? Is your goal to reduce your grocery bill by $30 a week? Maybe you have dreams of running a tiny but mighty farm yourself and want to sell to consumers down the road.

Regardless of your goals, creating a strategy will make or break your production. Without a clear vision of how much you want to preserve for the season, use fresh in the kitchen, or sell, you will have no idea how many seeds you need to sow in the ground or tomatoes you need to start in the greenhouse.

So, grab that pen and notepad, and let's get busy.

- What are your goals for food production?

- How much time do you have to dedicate to the preservation of your food?

- Do you want to participate in farmer's markets?

- How much are you spending on groceries now, and what can you do to eliminate some of those expenses?

Evaluating your answers to the questions listed above will be a good starting point for nailing down a strategy for yourself and your family, and deciding which crops you will grow to help achieve your goals.

Tiny tip

Here are a few simple ways to preserve your food.

- Freezing

- Dehydrating

- Freeze drying

- Canning (water bath and pressure canning)

- Fermenting

- Drying

When planning what to grow for the season, it is best to think through how you intend to preserve your harvest.

'Carmen' peppers will start green and turn red when they are ripe. They taste great grilled, roasted, in salads, or eaten raw with hummus.

High-quality crops

Early in my gardening adventures, I learned the value of figuring out what my family enjoyed eating. I spent countless summers growing vegetables in my garden that my children never touched. I did not have a plan—and it showed. I grew foods that we never bought at the farmer's market or cooked regularly. I soon learned that that needed to change.

Examine what foods you eat the most and grow those things first. Try not to become overwhelmed by the unknowns of the garden and take on each season hopeful of the abundance your gardens will produce for your family. Soon enough, you will have countertops overflowing with tomatoes that need to be preserved and baskets of veggies placed around your kitchen. What a glorious, rewarding thing that is!

But, to achieve that abundance, there is one crucial thing you need—high-quality crops. When you invest in a high-quality seed that serves multiple purposes, you start to unlock a domino effect. Plant crops that will add fertilizer

The cucumbers will climb to the top of the jute twine, then climb over the pipe and "umbrella" out. It really is a beauty to see so many cucumbers produced.

This is a good reminder to evaluate your goals often. If you had the goal of reducing your grocery budget, growing food you eat weekly is crucial. If preservation is your goal, then the crop you grow to store might be different. Keep your goals written down and nearby where you can reference them often.

and nutrients back into your soil, ones that are beneficial for the next crop going into the same space. The idea is to work smarter, not harder.

We can plant intentionally by choosing crops that are high producers and meet our growing needs best. Here's an example: You can plant a hybrid cucumber in a greenhouse in place of a standard seed, and it will produce not only enough for your family but also for the consumer market in less amount of time. You will have more uniform fruit and plants that handle stress better. Why is this? Because these are varieties that are bred for production purposes.

What is a hybrid?

A hybrid plant is the result of intentionally cross-pollinating two different varieties of a plant species. By doing this, you are hoping to combine the best traits and qualities of each parent plant into the offspring.

When thinking through quality crops, consider high-producing, multi-purpose crops. How can both your family and the market you are trying to reach benefit from your crops?

My favorite high-producing crops:

- 'Excelsior' organic (F1) cucumber

- 'Rainbow' (F1) carrot

- 'Grand Marshall' (F1) tomato

- 'Sprinter' organic (F1) bell pepper

- 'Rebelski' (F1) tomato

- 'Carmen' organic (F1) Corno di Toro Pepper

Hybrids sometimes get a bad rap, but these stunning 'Sakura' tomatoes are a showstopper on our farm. If your goal is production, hybrids are an excellent addition to any garden.

One aspect of growing is for high production, another aspect is for flavor and cooking which we will explore later in this chapter.

Layout

Once you've established the high-quality crops you intend to grow, you will need to evaluate the layout of your space. In chapter 5, we'll discuss the different styles of gardening.

The layout for your garden can prove challenging. I like to look at it from a static standpoint but also from a production state of mind—evaluating the planting process, seeding, transplanting, cultivating, harvesting, etc. All of these things play an important role in the layout of your gardens.

Planting schedules

Planting schedules are paramount for the home or market gardener. Thankfully I live in a growing zone where we have a very long growing season, but some of you may not be that fortunate. In that case, you will need to utilize the time you have to max out production and in on your planting schedules.

Tiny tip

The upside of having a longer growing season is that you are able to grow for a more extended period of time and into the fall and winter without spending more on row covers or even high tunnels. If your winters are milder than some climates, you are able to plant earlier in the season as well.

Let's back up a few steps. You will need to know the harvest date for each crop you are growing because that will determine when you need to start seeds indoors or direct seed them outdoors. Other factors will be whether you're growing in a tunnel or greenhouse, the time of year you are producing, the weather, the day length, and the types of crops you are planning to grow.

> **Fully grasping planting schedules was and still is the hardest job on our farm for me.**

Speaking from experience, fully grasping planting schedules was and still is the hardest job on our farm for me. You don't have to be a pro to get the job done, you just have to keep showing up. Be willing to learn and improve with each season. To help streamline this, I have created a checklist of a few common crops and their planting schedules.

Early or late plantings

Familiarize yourself with what hardiness zone you are growing in and your estimated first and last frost dates. If you aren't familiar with these, a quick internet search should offer some clarity. So will talking with fellow gardeners. I can almost guarantee that most gardeners in your area will know when to start planting and will be happy to pass on that information.

Once you've established the seeds you intend to plant and the timing of them, you will be able to start them early indoors or in a greenhouse, or you can direct seed them later when you are clear of spring frost.

Many crops can be direct seeded, but in order to get a head start on crops that take longer to mature, like brassicas, tomatoes, and peppers, starting indoors or in a greenhouse will ensure you receive yield.

You will hear me preach production throughout this book, but if you don't find what sparks joy throughout the process, it is all in vain. When considering layouts for your gardens think about creating a space that will not only be productive but create a space you want to keep coming back to.

Okra can be enjoyed many ways. Our most preferred way to eat it is freeze dried or dehydrated with some sea salt sprinkled on top for a nice summer snack.

Grow for flavor

Take heart. You can grow food (and a lot of it at that). Create a plan, buy some seeds (things you know you and your family will eat), and dive into planting. And don't stop there. One of the best parts of the garden—the abundance—is experiencing it together, with your family.

Truly make this food growing endeavor a family affair, and the best part, when you begin growing food with your family, they will begin to express their wants and desires, and you can grow food unique to their flavor preferences.

Growing food for flavor was not always something I emphasized. When you grow up buying food from the grocery store, you become accustomed to the mediocre produce you'll find

lined up on the shelves. The flavor isn't fresh. The herbs aren't as fragrant as the ones you can have if you were to walk out your back door and harvest them. You can tell the tomatoes were harvested under-ripe so they could make their way across the miles to the grocery store.

Grocery store food lacks freshness, and when it's grown out of season, harvested early, and packed in cold storage for months— everything changes.

Once I experienced how different fruits and vegetables tasted when grown in season and harvested at the peak time, that was it for me. I could no longer buy green beans and strawberries in the winter; these were seasonal crops that

CULTIVATING FAMILY

One of my favorite things about growing new varieties is the look on my family's faces when they've experienced a unique flavor for the first time. This is especially true when it comes to my children. I will never forget when Ivy June tried a tomato that tasted like pineapple for the first time or tasted a Mexican sour gherkin. I encourage you to try new varieties that broaden your family's taste palette. I was a bit intimidated to do this at first, but once I dove in, I was amazed at how much my family enjoyed the adventure of trying something new. They continue to be curious about what unusual flavors they will get to experience and never turn away an opportunity to try something new from the garden. It's a win-win.

My family is my driving factor behind growing food. I love offering new food to them, providing a level of sustainability to our family, and ultimately exposing them to the process of how food is grown, tended to, and enjoyed.

> **Once I experienced how different fruits and vegetables tasted when grown in season and harvested at the peak time, that was it for me. I could no longer buy green beans and strawberries in the winter.**

should be enjoyed within their appropriate local season.

After a few years of learning to grow with the seasons, I figured out how to eat with the seasons. Now butternut squashes are only a luxury for fall and winter, fresh salad is a spring staple, and okra is meant to be eaten on those hot summer days. Earlier, we discussed growing with production in mind, now, we are going to shift and discuss growing with flavor in mind. How you do this isn't always with using hybrid varieties that produce the maximum amount of yield, most times it is from heirloom seeds that were created with flavor in mind.

A few years ago I started to notice a shift in how people talked about heirlooms, grew heirlooms, and even searched for heirlooms. Now, almost any gardener you talk to, new or old, is aware of what an heirloom is. It has undoubtedly even become "a thing" to grow heirloom tomatoes, and I'd be the first one to admit, that there is something special about the unique colors, funky shapes, and variations of flavors.

The younger generations are certainly becoming aware of the history of these special seeds that have been passed down through the years.

If you want to explore cooking flavorful food from the abundance of your backyard, heirlooms should definitely be on your list of crop considerations. With thousands of options for heirloom varieties, you shouldn't have an issue finding a couple that stands out to you. Here are a few of our tried-and-true favorites.

My favorite heirloom crops:

- 'Dr Wyche's' heirloom tomato
- 'Black From Tula' tomato
- 'Dragon Tongue' bush beans
- 'Zucchino Rampicante' squash
- 'Shishito' peppers
- 'Blueberry Cherry' tomato

There are endless reasons to grow your food. The list goes on and on from production to flavor, sustainability, and togetherness with your loved ones.

I want you to leave this chapter having a clear idea of the goals, vision, and purpose behind why you are growing food. If you are still unclear about your vision, reach out to a friend, or ask for a mentor who's walked these steps before you to link arms and help guide you through some of the topics discussed. Having an outside perspective can prove extremely beneficial.

Remember to write down your growing zone, work on establishing crop plans, and the varieties you want to grow, and most importantly, in the midst of all the hard work and sweat, don't forget the purpose behind why you wanted to start your farm to begin with.

If you shift your focus to emphasizing soil
health, then the abundant food will come.
But without prioritizing healthy soil,
we will not be successful growers.

CHAPTER FOUR

Soil: The Health & Longevity of Your Farm

You know those moments years ago that you sometimes look back on—memories that are so vividly clear it's as though they just took place? That's what this story is, a memory of a very young girl who wanted to farm but had no idea what that even really meant. She was determined, headstrong (others called her stubborn), and persistent. Her garden was flawed, her vegetables would often die from a pest she couldn't even pronounce, and her soil...well, at the time, she just referred to it as good ol' dirt.

—☀—

One day though, while talking with her papaw, he turned to her and said, "Sugar, dirt ain't what's growing all my produce, the soil is."

I did not understand what he meant at the time, but now I fully understand how different dirt and soil are. Dirt does not host an entire ecosystem of microbes and microorganisms. Dirt is just that, dirt. It's not life-giving, nor is it alive and full of beneficial organisms. It really is just plain ol' dirt. The soil, on the other hand, is quite intricate. Healthy soil is very much alive and hosts millions of microorganisms. It's beautiful and produces abundant fruits and vegetables in your garden. It is composed of many elements that have been around since the Earth was created. The takeaway from this is that soil supports life in your plants and is a self-sustaining ecosystem that we need to protect for the longevity of our farms.

By the end of this chapter, I want you, the growing gardener, to feel hopeful that you too can not only understand soil health but also turn plain ol' dirt into healthy, life-giving soil for your farm.

Soil supports life in your plants and is a self-sustaining ecosystem that we need to protect for the longevity of our farm.

One way you'll know if you have healthy soil is when you start to see life in it. Worms crawling around, spiders, rollie pollies; these are all welcomed in my garden and should be in yours, too.

Soil is one of those things that, more than likely, is not wooing you to the garden. Few of us daydream about how beautifully captivating that smelly, dirty stuff is that we dig into each season. But, alas, soil is the backbone and longevity of your farm. I hope you will embark on this chapter with anticipation of how you can view your soil as the asset it truly is.

What we plant in the soil of contemplation, we shall reap in the harvest of action.
—Meister Eckhart

Why soil health matters

Everything begins in the soil. You would not or could not have a garden without it. (Unless you are growing using the methods of hydroponics or aquaponics, neither of which we will be discussing in this book.)

So, with that in mind, you get what you give. The more you pour into and add to your soil to improve the quality, the more you will get out of it, and the healthier your plants will be. Although it may not seem like the most exciting thing to dedicate resources to, it is valuable and necessary for your farm.

What is soil?

Soil, by definition, is a mixture of organic matter, minerals, gasses, liquids, and organisms that, together, support life.

Every time, the end of this definition resonates deeply with me—"...that, together, support life." Your soil structure and quality are the life support of your garden and farm. I'm not sure about you, but by knowing this, I think it is a safe assumption that we should all be putting more emphasis and time into improving our soil's health.

It is important to only add nutrients to your soil that you know it's been depleted of. We cover this later in the chapter about the importance of a soil analysis. Here I am amending with blood meal and feather meal.

Digging deeper

High soil quality does not happen overnight. It takes time. You will need to provide adequate nutrients and put organic matter back into your soil, season after season. You do that by feeding your soil precisely what has been depleted during the prior season. Compost is an excellent way to feed your soil. If you are striving to grow an intensive garden, you will be pulling many nutrients away from the soil, and by adding compost, you can build the soil fertility back up and enhance its structure.

How do you know what your soil has been depleted of? By conducting a soil test. A soil test can seem complicated to a beginner, but I cannot tell you how valuable this information will be for you as you start growing your food, so let me help you break it down.

To take a soil test go out to your garden beds and collect soil from multiple parts of the beds (in-ground or above-ground). I find it easiest to use baggies so I can write on the front which beds the soil came from. They're also simple to transport.

You can then take that soil sample to your local university or governmental agricultural facilities. There are also independent soil testing laboratories you can send the sample to, which is what we do with ours. Typically, they test for various nutrients,

like phosphorus and potassium, and your soil's pH level (the level of acidity or alkalinity), but you can request that your micronutrients and organic matter levels to be tested, too. Usually, that comes at an additional cost.

When you receive the results, they will be detailed and may seem overwhelming. But, keep the faith, and remember there is a purpose that will allow your farm to flourish and thrive in the upcoming season!

I would encourage you to call the university or the laboratory you submitted the sample to and have them walk you through your results and what exactly they mean. Ask about the organic amendments they suggest. When I did this, they were more than willing to take the time to review our results.

Soil test checklist:

1. Gather your supplies: baggies (or whatever you choose to collect your soil in), marker, and a trowel or shovel

2. It is important to get soil from several locations in your garden. I do a bag for each of my north, south, east, and west beds, one for the high

Soil samples are easy to collect and aren't nearly as intimidating as they may seem. I recommend testing your soil each season and amending your beds appropriately.

It is important to label your containers or baggies. On each baggie I write the location (raised bed, high tunnel, cottage garden). Typically I am pulling samples from the north, south, east, and west parts of the beds, which is also labeled.

tunnel, one for the cottage garden, and one for the raised beds. In any area you plan to grow food, I would send off a sample of soil to be tested.

3. Ensure you provide enough soil for adequate testing. I send 1–2 cups (250–500 ml) of soil from each of the various beds (all labelled accordingly). I take the soil from the root zone, at a 4–6 inch (10–15 cm) depth which is typically the length of a trowel.

Feeding your soil

Soil is an essential asset on your farm. If you have been growing your food for any period of time, you know there can be multiple ways to do something. It is not a one-size-fits-all sort of thing. What one gardener's soil needs can be and will most likely be different from another. What I can provide you with, though, are the principles to ask yourself when building your soil quality. A few things can turn your infertile soil into rich, fertile soil. Usually, these are minerals, compost, and organic matter. When you provide your soil with enough of each of these components, you will notice that it starts to excel. Throughout this chapter, you will hear the terms organic matter and compost. Here is what I am referring to.

Compost From *Wikipedia*, "Compost is a mixture of ingredients used to fertilize and improve the soil. It is commonly prepared by decomposing plant and food waste and organic recycling materials. The resulting mixture is rich in plant nutrients and beneficial organisms, such as worms and fungal mycelium."

Organic matter From *Wikipedia*, "Organic matter, organic material, or natural organic matter refers to the large source of carbon-based compounds found within natural and engineered, terrestrial, and aquatic environments. It is matter composed of organic compounds that have come from the feces and remains of organisms such as plants and animals."

These two components are essential because of their role in your soil's life.

Organic matter of good quality will be the foundation for the microbiology in your soil. You could look at it as the supply of nutrients for your plants. It reduces compaction, provides soil structure, increases the nutrients in your soil, allows moisture retention, and increases water and air infiltration into your soil. I like to look at it as the driving factor that makes all the other components work harmoniously.

When you have healthy soil, it produces healthy plants. If you start to notice deficiencies in your plants, lack of growth, etc., always examine your soil first.

SOIL STRUCTURE

A soil's structure is the way its particles aggregate together. Good soil structure is key to allowing air and water to move through your soil, which is vital to the growth of your plants. Without structure in your soil, you will have issues like erosion, poor drainage, and locked-up nutrients causing issues in your plants.

Compost

Organic matter is beneficial for your soil, and compost is a great source of it. Compost is an amendment that can add nutrients and life back into your soil, and the best part is you can play a role in making your own compost right in your backyard. Yes, there are a few steps involved, but being able to build quality soil over time using homemade compost will save you money and time in the long run.

If you aren't familiar with composting, here is a breakdown. Composting is a microbial process that converts plant materials, such as grass clippings and leaves, into a usable organic soil amendment.

Compost has been used for centuries to increase organic soil matter, improve the physical properties of the soil, and help supply nutrients for successful plant growth.

One of the biggest benefits of composting is that it reduces the need for chemical fertilizers. When you start building a compost pile, the beneficial bacteria and fungi begin to break down the materials in the pile and create humus, which is a rich, nutrient-filled material that is extremely beneficial for your plants and your garden soil.

Composting is a microbial process that converts plant materials, such as grass clippings and leaves, into a usable organic soil amendment.

As we dive into the science of composting, I don't want you to feel overwhelmed. It takes years to fully understand the creation and use of compost. Focus on a few things that make sense to you and seem within reach. Then, once you've fully understood the working parts of those things, begin to add a few more components, and so on and so forth.

The science of compost

Microorganisms are vital to the composting process. One key to effective composting is to create an ideal environment for the microorganisms to thrive in that has warm temperatures, nutrients, moisture, and plenty of oxygen.

There are three main stages in a composting cycle that allow it to thrive and succeed.

Stage 1 The first stage is typically only a few days long during which mesophilic microorganisms thrive, the temperature of your compost is between 68°F–113°F (20°C–45°C).

Stage 2 The second stage can last anywhere from a few days to a few months. The thermophilic microbes begin working to break down the organic materials into smaller pieces. The higher the temperature of your compost, the faster it will begin to break down the proteins, fats, and carbohydrates.

During the second stage, your temperatures will begin to increase, it is crucial to keep an eye on this so your compost doesn't become "too hot" and kill off the beneficial microorganisms.

When you turn your compost pile, this allows aeration to take place, helping to keep your soil temperature lower.

Stage 3 The final stage will typically last for several months and begins when the thermophilic microorganisms use up all the available nutrients. Typically, this stage is when your compost temperature will begin dropping enough to allow the microorganisms to regain control of the compost pile and finish breaking down the remaining organic matter into usable humus.

A few things to note when creating compost at home are about what you can and cannot add to your compost (see the Tiny Tip on page 74.) and how to maintain your compost pile for the best soil results.

When your soil is alive and fertile, your plants will be bursting with life. Swiss chard thrives in healthy soil. It can be grown in succession plantings through most of the growing season and even into the fall and winter, depending on where you live.

How to build a compost bin in your backyard

1. Figure out your compost area or compost bin placement. Be mindful when choosing the location (think long-term) and allow yourself enough space to be able to get equipment in the area to occasionally turn your compost over. If it's a big compost pile, that may mean room for a tractor.

2. Gather building materials for the bin. This can be pallets, tin, or cinder blocks. We use sheets of "snow fence" and have a designated area for the compost, allowing room for our chickens to still have access to scratch around in the pile and help break down the compost components.

3. Build the bin. Be sure to add enough support to brace your pallets or other bin-building materials.

4. Begin adding material to your compost and allow the breakdown process to happen. The smaller the material size that is added to your compost, the easier and quicker it will break down.

Tiny tip

Add things to your compost such as: fruits and vegetables, eggshells, animal manure, coffee grounds, hay and straw, leaves, tea bags, and wood chips.

Avoid adding citrus rinds, meat, dairy products, or diseased plants.

It is good practice to turn your compost weekly, either with a tractor or shovels, depending on its size. The more you turn your pile, the faster your compost will begin to break down and turn your scraps into finished compost.

What to compost

When thinking about what to add to your compost pile, think about a balance of green and brown materials. A compost pile needs both to thrive. Brown materials are high carbon components such as dry leaves, wood chips, straw, sawdust, or cardboard. Green materials are high nitrogen components such as food scraps, plants, grass clippings, and manure.

This is a basic list as a reference, there are plenty of things you can and cannot add to your compost pile. If you are unsure if something is safe to add to your compost pile, a quick internet search should help.

When it comes to your farm, there are many
things you can pour your time and resources
into, but focusing on soil quality will always
yield the best results. Growing high-quality
food right from your backyard will be worth the
time you spend getting your soil right.

When it comes to compost, you will likely
spend the next several years growing and evolving
your knowledge, it's okay to mess up and study
the lessons again. Much like everything else in
the garden, there is grace in this area. Soil isn't a
quick fix; it takes a long time of being intentional
and pouring nutrients back into your soil for it to
produce bountiful crops. A compost pile is the
same way. You won't have "gardener's gold" in a
few days. It will be months of diligently adding

materials, turning, and building the quality, but if
you stay the course, over time you will pat yourself
on the back for a job well done. Compost is just
one way to add nutrients back into your soil, and
if you aren't at a place to create a compost pile
yourself, you can easily source compost locally or
choose other amendment options.

Nothing will put a smile on my face faster than seeing bees and butterflies throughout my garden. Being intentional with a space for them will prove beneficial for your crops.

Diversity in the soil

When establishing healthy soil on your farm, diversity is a crucial component. Plants will feed the beneficial microbes in your soil. This happens through photosynthesis. Plants use the energy from the sun to turn carbon dioxide and water into carbohydrates. That's pretty cool, right? It gets better. From there, the carbohydrates start to feed the plant, which allows it to start growing roots, stems, leaves, etc.

Plants can also disperse those carbohydrates directly to the soil through their roots, which feed the bacteria, fungi, and other beneficial microbes in the surrounding soil. You may have heard of "beneficial microbes" before. When you've created healthy, life-giving soil, you'll have a large and diverse mix of these microbes. While we're discussing diversifying, I'll mention that different plants feed different microbes. The more you diversify what you grow on your farm and in your garden, the more diverse your microbes will be in your soil. You can achieve this by rotating what you grow in your garden beds from season to season— mixing various flowers with your crop rotation of vegetables. When you do that, you also create a friendly environment to attract pollinators, which you want in your garden, in addition to creating healthier soil.

> **"Fertility of the soil is the future of civilization."**
> —Albert Howard

Working in harmony

We all want thriving gardens that we can feel proud to show our friends and family. We want to walk around our farm and see life bursting from the garden beds. But we only achieve this by understanding the balance our soil needs and learning how to create harmony within a healthy garden ecosystem.

I was talking with my mentor one day and he threw around the phrase "working in harmony with nature." I understood what he was saying, but I wasn't aware of how to achieve this on my farm. If you feel much like I did on that today, then keep reading, friend; we are about to get to the good part.

How we encourage our farm to work in harmony with nature will be slightly different for each of us. This is based on many things, including your values, your goals, and your farm itself.

When I think of working in harmony with nature, I envision planting more for not only myself, but for the other living things around me, like the bees, butterflies, and birds. I also envision not adding synthetic chemical fertilizers and pesticides that might harm or kill those beneficial creatures in my garden.

Working with harmony makes me think of using animals on our farm to prepare the ground for garden space, instead of relying on machinery and tilling that will break down the soil, rather than bring life to it.

What can we do on our farms to keep our integrity intact and rely on the nature around us to help achieve our goals?

I'm not sure any of us will ever reach the ultimate goal of success, either through our soil management or by working harmoniously with nature. Soil should be something we are continuing to improve, season after season, year after year. It will improve, and you will start to see life literally bursting from your soil. This is how you know you are on the right path. You will also see bees pollinating and plants thriving. This is how you know that you've cultivated a healthy ecosystem. The goal is to keep life thriving and build upon that. If you are committed to continuously adding to your soil and to the integrity of your farm, your gardens will continue to provide you with abundance and beauty for years to come.

Incorporate trap crops in your garden to combat pests and eliminate the need for chemical inputs.

Balance

Understanding soil needs is really about learning how to balance the many components of soil. Look at soil as the foundational building block to a successful garden. The first step is to reference the soil test (covered earlier in this chapter). Your soil analysis will tell you what your soil needs and how to start achieving that balance. When we add nutrients to our soil, we need to ensure that we properly go about it. Your soil is made up of physical, chemical, and biological properties. These are important for root growth and microbial activity. You also need components like water and air for soil to be balanced, which we will touch on next.

When adding essential nutrients, it can be easy to overload your soil with one nutrient, which could, in turn, make another nutrient deficient by not allowing the plant to absorb what it needs.

Your soil pH is also a crucial part of knowing your soil needs. Your ideal soil pH will depend on what crops you are growing, but if you add too much fertilizer, you can inadvertently change the pH of your soil in the process. Doing this could mean you are changing the soil conditions from those your plants like to grow in, to those they do not want to grow in. It is good practice to test your pH frequently to ensure your plants have the optimum pH for their needs. If you continue with the preparation of rotating crops and diversifying your plantings, a general soil pH of 6.5 is ideal. Many factors will come into play when finding your soil's optimum balance. Your region and growing zones play a role in this, too, and as mentioned earlier, so does the variety of crops you are growing.

Air

All right, friends, we are about to get technical here. Buckle in and think about all the tools you will have in your tool belt once we are done. First up, air. The air we breathe is a mixture of gasses made up of 78% nitrogen, 21% oxygen, and a small percent of water vapors and other components. One thing I admire about our world is how it is all interwoven. Did you know that the oxygen found in the air we breathe is there because living things like plants have produced that oxygen for over 2 billion years through photosynthesis? The carbon dioxide you and I breathe out is required for plants to trap their energy from the sun and turn it into food through photosynthesis, releasing oxygen in the process. Pretty fascinating, right?

When referring to the air in your soil, it is present in the pore space and is usually called the soil air. Soil air will support the life of your plants and other organisms in your soil. It helps with the growth of the aerobic microorganisms there.

Often, we will hear the term "well-aerated soil." This means things are functioning correctly. It means there is enough air in the pore space. When your soil is well aerated, it means gasses are available for your plants to use, and that there are macroorganisms or microorganisms present.

One reason you may have poorly aerated soil is the presence of excessive soil moisture. The excess

When your soil is functioning properly your plants thrive. Granted, this doesn't mean you won't ever have a pest in your garden or a diseased plant to rip out, but it will be far less often when you have the proper nutrients in your soil.

A general rule we follow on our farm is to amend the soil before each planting and only to amend what it is deficient in. You can add too much of a good thing. Remember to reference your soil analysis.

water can occur for various reasons, including overwatering, poor drainage, or poor soil structure that doesn't have large enough pore spaces for the water to freely drain away. All of these can affect your plants and the beneficial organisms in your soil.

When you have adequate aeration, it begins to change the physical properties within your soil. You'll notice improvements in structure (light and fluffy), density, and porosity which will be important later.

Water

Water is a critical soil component, and when it's contained in the soil, it is referred to as soil moisture. The water is held within the pores of the soil wherever there is an absence of air. There

should be a balance between soil water and air. Both are a crucial part of the soil and are involved in the availability of nutrients in the ground. It is essential to ensure you have the proper moisture levels. When there is ample soil moisture, it ensures certain nutrients are readily available for your plants to take up and use as necessary.

Water is the medium from which plants assimilate certain plant nutrients. Soil water contains dissolved organic and inorganic substances, and it transports certain mobile nutrients, such as nitrogen, sulfur, calcium, and magnesium, to the plants' roots for absorption. We cannot have healthy, thriving plants without adequate water in our soil.

Soil water is essential for photosynthesis. In addition, soil moisture helps regulate the soil temperature. It is also required by the microorganisms in your soil that require it for metabolic processes.

Soil structure

Throughout this book, you will hear me referencing your soil structure because this is something I believe you should become well acquainted with. Building, improving, and maintaining soil structure is imperative. Your soil's structure refers to the way its particles are grouped into aggregates. The structure is the shape your soil takes based on its chemical, physical, and biological properties.

When you have well-structured soil, it will be crumbly and have a good amount of pore space that allows water and air to move around and support healthy roots. Good structure is also essential to improve drainage and reduce the risk of soil erosion caused by excess runoff. When the soil is not fertile and compaction starts to set in, its inadequate structure interferes with many other moving parts in your soil, which will affect how everything else flows through and operates within.

Little life, big impact

I always welcome those slimy little animals that move across my hands as I'm digging in my garden bed. Naturally, I jump, thinking that the small animal might be something scary, but there is nothing dangerous about earthworms and other soil critters. Their activity in the soil offers a range of benefits. Earthworms eat decaying leaves, rotting plants, compost, and even the soil itself. This is mixed with the juices in their digestive tract and the enzymes in their stomach, and they return it to the soil through their castings (yes, I mean their poop).

Worms are fascinating. Their little lives have had such a significant impact on our farm. The earthworms help create humus, which is that dark rich soil that holds nutrients in place and encourages plant growth. They also enable proper soil structure. Their burrows open up the ground, allowing aeration and soil drainage channels to form.

When the earthworm's castings are dispersed throughout your soil, they will be rich in N-P-K (nitrogen, phosphorus, and potassium). These are vital minerals needed to maximize plant growth. When the worms burrow through the soil, they create holes that allow air to reach plant roots. The same tunnels also allow rain and irrigation water to infiltrate the soil. And, one of my favorite benefits of the earthworm is the castings they leave behind. They contain 5 to 11 times the amount of N-P-K as the materials they've ingested. Talk about a return. So next time you see those little fellows in your garden beds, think about the extraordinary capabilities they have and how beneficial they are to the life of your soil.

If you notice large clumps of dirt that won't break down in the soil, or perhaps you never see life in your soil, that is a good indication that something is off. A healthy ecosystem in your soil will always produce life. Be on the lookout for loose, dark, rich soil crawling with life.

Whether you plan on growing vegetables, flowers, or established perennial berries and fruit trees, soil is vital to the growth and success of any crop you grow on your farm.

You can do this!

I understand this is a lot of information, but soil care becomes easier each season. Eventually, you will not need help to break down your soil test results for you. You will read it and immediately know how to add nutrients back to your beds and improve the structure. And, when you hear words like organic matter and compost, you will immediately think about how they are feeding all those beneficial microbes in your soil. As your knowledge deepens, you will become more confident in your gardening journey.

Think about soil as the backbone of your farm. It is the foundation you will use to produce food, preserve food, and even sell food. It's well worth the time put into it on the front end for the longevity of your farm on the back end.

THINGS TO REMEMBER
Soil = the life of your farm.

You give what you get. Spend quality time investing in understanding and knowing how to improve your soil structure and quality.

It takes time. You will not become an expert in soil overnight, and that is okay. This is where we get to do this fun thing called growing. We are growing knowledge and comfort levels, right along with growing food. You are gaining experience through the act of doing, and you will feel confident in understanding your soil needs in no time at all.

List of organic amendments for feeding your soil

Feather meal A byproduct of processing poultry. Feather meal is made from poultry feathers by partially grinding them under heat and pressure and then drying them. When adding feather meal to your garden as a nitrogen source, it needs to be blended into the soil to start the decomposition process to make the nitrogenous compounds available to the plants for absorption.

Bone meal A mixture of finely and coarsely ground animal bones that were first steamed, then dried and ground. Bone meal is an excellent source of calcium and phosphorus.

Blood meal A derived product made from blood. Typically, the blood comes from livestock production and would otherwise be a waste product. Blood meal is used as a high nitrogen organic fertilizer. Blood meal is different from bone meal because it contains a high amount of nitrogen, while bone meal contains phosphorus.

Kelp meal Made from ocean seaweed that is dried and ground up into a fine powder. Kelp contains small amounts of all the N-P-K nutrients, which is why it is excellent for your plants, and it is a fertilizer that is hard to overuse. While it carries a small amount of nitrogen, phosphorus, and potassium, it has over 60 trace minerals or micronutrients. These include calcium, magnesium, sulfur, copper, iron, and zinc, to name a few.

Compost Compost is decomposing organic materials. It contains macronutrients and micronutrients (this is usually lacking in most synthetic fertilizers). Compost is slow release and can continue feeding your soil for months. It acts as a buffer for the soil and can neutralize both acidic and alkaline soils. It will help bring your pH levels to the optimum range for plants to absorb nutrients.

Wood ash Wood ash contains phosphorus, calcium, magnesium, potassium, and other essential nutrients.

Wood chips Wood chips are small pieces of wood. They are made by cutting or chipping larger trees, branches, stumps, or various other wood materials into smaller pieces, making them easier to break down. They are used as an organic mulching option for the garden.

Soil Amendments:
What They Are and How They Affect Your Soil

About N-P-K		
N=Nitrogen		
P=Phosphorus		
K=Potassium		
Nitrogen Sources (N)	Formulation	N-P-K Ratio
Organic Amendments		
Feather meal	Slow release	12-0-0
Blood meal	Slow release	13-1-1
Manure	Highly variable	
Coffee grounds		2-0-0
Fish emulsion		4-1-1
Commercial Chemicals		
Ammonium nitrate		32-0-0
Urea		46-0-0
Ammonium sulphate		21-0-0
Phosphorus Sources (P)		
Organic Amendments		
Food scraps typically added to compost		
Bone meal	Slow release	4-15-1
Manure	Highly variable	
Pelleted chicken manure		3-3-3
Commercial Chemicals		
Diammonium phosphate (DAP)		18-46-0
Triple superphosphate		0-46-0

Potassium Sources (K)	Formulation	N-P-K Ratio
Organic Amendments		
Kelp meal		1-0-4
Wood ash		0-2-10
Potash		0-0-60
Greensand	Slow release	0-0-3
Banana peels typically added to compost		
Commercial Chemicals		
Potassium chloride (muriate of potash)		0-0-60
Potassium sulfate		0-0-50

List of organic and inorganic soil conditioners to improve soil structure

Parboiled rice hulls A renewable byproduct of the rice milling process. The hulls are sterilized with steam and then dried. Rice hulls behave similarly to perlite, improving aeration and drainage. Reduces the heaviness of clay soils.

Compost (See page 83)

Ramial chipped wood This chipped wood product is made mostly from medium to small-sized branches and contains a significant amount of pulverized leaves. Compared to regular wood chips, bark mulch, or shredded hardwood mulch, ramial chipped wood has a significantly higher nitrogen to carbon ratio. Additionally, it also contains significant amounts of trace minerals.

Limestone Limestone can help remediate overly acidic soils while also adding calcium. Having pH-balanced soils mean essential nutrients are more readily available.

Bentonite clay Add this clay to sandy soils and it improves water and nutrient retention.

List of inorganic conditioners and mediums for potting soils

Perlite A volcanic glass that is composed of about 70 percent silicon dioxide. It has high water retention abilities. It is used to improve soil structure by providing better aeration and drainage.

Vermiculite A naturally occurring mineral that expands when exposed to extreme heat. It is common in the gardening world and used to increase water and nutrient absorption while aerating the soil.

Peat moss A fibrous material that forms when moss and other components decompose in peat bogs. It helps hold moisture and nutrients into your soil.

It has been years since I had that conversation with my papaw about dirt and soil. I look back on it now and just smile. Man, I sure didn't know what I was doing. Thankfully, I was willing to learn. What I once referred to as dirt, will never be referred to as that again. The soil I've spent years building on my past farm, and now the soil on my current farm, is the driving force behind everything we do. I can grab a handful and see worms crawling around. I can walk through my garden and feel the power of the Earth under my feet. I can harvest the fruits of my labor because I've been intentional with my soil needs and the life found within it. Never underestimate the power that's in your soil. Remember, that it is the lifeline for everything your farm produces. Give it the time and attention it deserves, and one day you'll have your own funny story to look back on and share.

Tiny tip
Tracking your soil's progress is very encouraging and a great motivator along your gardening journey. I recommend saving the analysis from your soil test every season to see the improvements that have taken place. This confidence boost will catapult you forward.

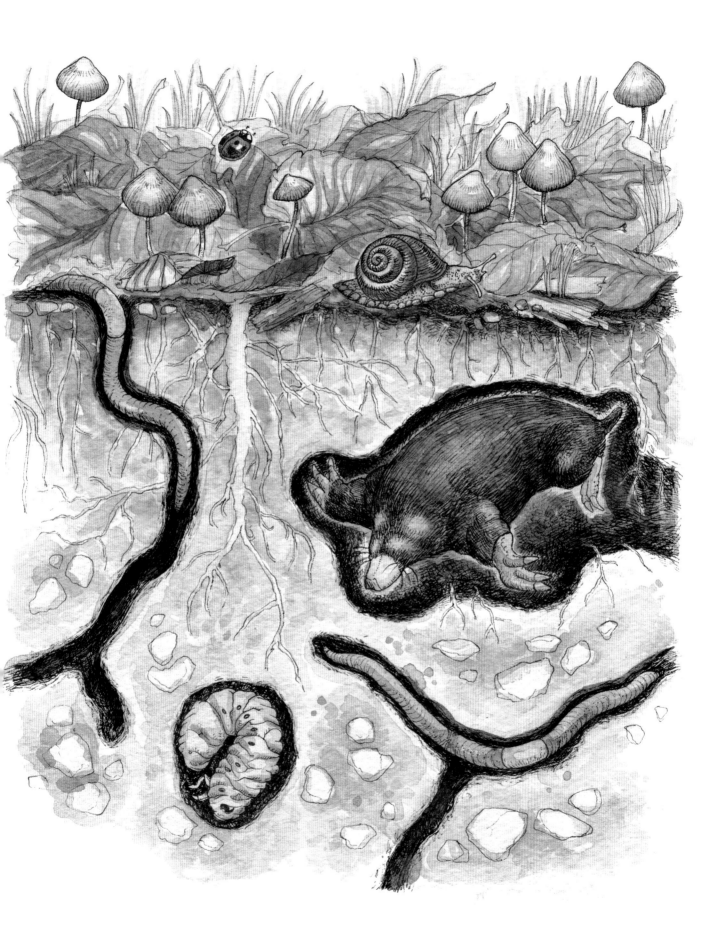

I like to think of gardening as having
guidelines, not rules.

CHAPTER FIVE

In-Ground, Raised Beds & Indoor Growing: Which is Right for You?

The options for garden styles and designs are endless. We can look around our farms or backyard space, dream up any possibility, and make it happen. That inspires me. I have the freedom to implement various gardening styles on my farm because I can. No one makes the rules for your garden except you. What works for one gardener may not work for another. That's where different styles come into play. I like to think of gardening as having guidelines, not rules. I've always been the type that didn't follow everyone else. Maybe that's why I enjoy having freedom in the garden. It is just me, the earth, and my wild imagination, and no one can hinder that creativity except me.

A lot of times, our vision of our garden's potential is stunted by a picture in our mind from our youth. Maybe your parents or grandparents had a traditional garden where they plowed the ground, raked the dirt into rows, hand seeded all their crops, and watered their garden like crazy. Maybe your only experience with gardening growing up consisted of a small herb bed right off the front porch of your home. My own garden vision from my youth was of the in-ground approach, with long orderly rows of the same variety. I learned quickly to take the things I had seen in my childhood as inspiration, but not let them be a stumbling block to moving forward. I wanted to push myself as a grower and try different techniques of gardening.

Use the garden as an opportunity to let your mind freely create. Create a space for production, but also create a space that is beautiful and full of life.

When you're trying something new, it is easy to immediately think about your return on investment. Remember, you're investing in yourself as the gardener as much as you're investing in the garden itself.

Before we dive into this chapter, keep in mind that not all of the different types of gardens discussed here will resonate with you. That's okay. But I do encourage you to lean in, challenge yourself, and choose the style that best suits your current lifestyle.

Allow yourself to think differently about your gardening space. Understand that you do not have to be married to one idea. You can grow food using any gardening style, some not even mentioned in this book. That's the cool thing about nature; it prevails. Food can grow from rugged, rocky, less than ideal growing conditions and still feed your family. Food can grow in the ground, in raised beds, in containers, and even on your kitchen counter. Your gardening styles will likely change through the years, but food will still be abundant if you

I like to think of gardening as having guidelines, not rules.

If you are starting out, I encourage you to start with a few grow bags or containers, and once your experience grows, add on raised beds and in-ground growing spaces.

GreenStalk Planters are not only wonderful if you are limited on space, but they add depth and textures to your garden and are perfect for keeping those leafy greens nice and clean.

CULTIVATING SELF

I love walking out my front door and seeing a cottage garden filled with textures and swirly rock beds. My raised-bed garden houses more than just vegetables and flowers. You will find a pergola lined with wind chimes and hanging lights. Out back by our in-ground space, we have a small bistro table and chairs where you'll find me in the afternoon enjoying a glass of lemonade. I have these spaces to come and rest, to feel at peace with what I've worked so diligently and hard for. Create spaces like these throughout your garden, take time to rest, and enjoy the journey.

If you're thinking of adding a pollinator patch, use this opportunity to grow what sparks joy. You can grow various varieties of zinnias, like this cactus one, daffodil and tulip bulbs in the spring, and patches and patches of sunflowers late into the summer. This is your chance to plant beautiful blooms that will draw you back to your garden every day.

implement the good practices learned throughout this book.

I want to set you up with the success you need to thrive on your farm. As we navigate through the various gardening styles in the coming pages, think about what you envision for your space. What will spark joy when you walk outside every day? Once you've figured out what that is, grow it in abundance.

On my farm, we demonstrate multiple gardening styles. We have a small plot focused on in-ground gardening, but my large garden and high tunnel are raised beds. In the cottage garden, you'll find containers and tower gardens. Understanding these different gardening styles will give you a better idea of which one is right for your small farm (and it might be all of them). I want to discuss these and provide some pros and cons to each style.

Tiny tip

When expanding your gardens, think about more than the style of the garden. Think about extending an invitation for nature to come and play a role in the health of your farm. A few ways to do that are by creating designated spaces for nature to inhabit. Plant native grasses and flowers for bees to come enjoy. Plant milkweed for the butterflies. Place pockets of natural habitats around your gardens to extend the invitation, and always plant enough for both you and the wildlife around you. This may seem like an extra step, but if you implement this tiny tip, I think you'll be pleasantly surprised by the quantity of life your gardens will support.

Gardening styles

There are several different ways to grow food, each with its own benefits and drawbacks. Before choosing which is best for you, there are some factors to consider. These considerations are the same for beginners creating their first garden or experienced gardeners intending to garden on a bigger scale. These are the things to think through before you start.

1. Evaluate your space.

2. Consider your budget.

3. Know your goals.

Raised beds

The term raised bed can be confusing. If you bring in a couple of inches of soil and build it up from the ground level, that technically is a raised bed. But when most of us think of raised beds, we *aren't* imagining a few inches of soil built up from the ground with zero structures around it. We are thinking of an aesthetic, framed bed made out of various materials filled with soil.

Let's discuss the different raised-bed options.

Permanent raised beds

These would be the permanent type of raised bed mentioned above. They involve building a structure around which the soil is held in place. This could also be considered a bordered raised bed.

The raised beds in this style of garden can be of any size and can be made out of any material of your choice—wood, tin, cinder blocks, rocks, etc. One benefit of creating a permanent raised bed in your garden space is the opportunity to use recycled materials (some of which may even come from on-site). All the raised beds in my garden were built by the previous owners using trees from our farm and tin they had collected over the years from various projects.

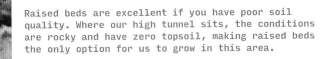

Raised beds are excellent if you have poor soil quality. Where our high tunnel sits, the conditions are rocky and have zero topsoil, making raised beds the only option for us to grow in this area.

One of my greatest responsibilities is to raise a generation that is connected to their food. I do this by creating a space for them. I always say yes when they ask to join me for mornings in the greenhouse or for harvesting in the raised beds.

Semi-permanent raised beds

Another option would be to build semi-permanent raised beds. These are several inches up from the ground level and have a couple of feet between the rows, but you are not using any material to border the bed and hold the soil in place. Since they are not framed, you can maximize the plantings in your space more efficiently and move the beds if need be. You would mound the soil to make this semi-permanent raised bed. This type of bed will still reduce the compaction of the soil and will stay in place for several seasons—or even several years if you'd like.

Temporary raised beds

Temporary raised beds consist of a few inches of soil built up from ground level and don't have a structure bordering the bed. They will allow you to grow plants in seasonal rotations. You may create these temporary beds and use them for a single season, then push the soil aside and plant a cover crop, putting more nutrients back into the soil. The following season, you'll build the soil up again, and continue alternating the harvests with cover crops. Often, you see large farms practicing this.

What is a cover crop?

A cover crop is a non-edible crop planted with the intention to not harvest it, but instead to ensure the soil isn't left bare. Cover crops reduce soil erosion, and once they break down, they improve soil fertility and quality. They also help reduce pest pressure and can be extremely beneficial for the pollinators and wildlife around you.

Traditional permanent raised beds are often the priciest of these options if you can't use recycled materials. If your goal is low cost and high efficiency, you might already see how this may not be a good fit, and we've not even done a deep dive into the cost of the soil yet. But before we get ahead of ourselves here, let's remember this: there is no right or wrong answer. As I mentioned, your space, your goals, and your budget should all be considered before making these decisions for your tiny farm. Here are some pros and cons to building raised beds.

Raised beds can vary is shape, size, design, and many other factors. Don't limit yourself to sticking with one style of raised-bed garden.

The pros and cons of raised beds

Pros of raised beds

- If you have poor soil quality or unusable ground, raised beds allow you to bring in fertile, rich soil. You can practice a "no-till" style.

- Raised beds protect the soil structure due to lack of disturbances like walking in the beds which can create compacted soil.

- You can cultivate and amend your beds from either side. This leads to reduced plant damage.

- With fertile, rich soil, your raised bed will have better drainage, especially in regions with clay soil.

- If you have sandy soil, the water retention will be better.

- Raised beds can be an excellent way to grow food if you have limited mobility or are disabled.

- Raised beds are manageable for smaller spaces, and if using the permanent, non-bordered raised bed approach, you can maximize your yields in a tiny space, creating efficient systems and crop rotations.

Cons of raised beds

- Of all the options, the bordered raised bed is undoubtedly the most costly. If you aspire to max out the production of your tiny farm and need to build multiple bordered raised beds, the cost will add up quickly. There will be the cost of materials to make the raised beds and the cost of soil to fill them. As mentioned earlier, you can use recycled materials, but this will be labor-intensive. If you choose to use wood from your property, you will need to cut the trees and make them usable for the beds. If you want to make your beds out of tin, stone, or other materials, you will still have to consider your time to collect enough of each item to build the structures. If aesthetics are important, your costs continue to rise.

- Weed management would be the same as practicing no-till gardening, but the no-till method (see below) is cheaper and more beneficial if you want to maximize yields in a small space.

- Raised beds tend to dry out quickly, the rate of which will depend on the material used to construct the beds as well as the composition of the soil in them.

- Since your raised beds will likely dry out faster, you will need to water more frequently, and if you don't have irrigation in place, this can become very time-consuming.

- Raised beds are not easily moved. There isn't a lot of flexibility when it comes to relocating your raised beds. So, when you decide on the location and style of your raised bed garden, you will need to ensure it is where you want it to be for the long-term.

- A bordered raised bed doesn't allow as much flexibility in your farm's layout.

Raised beds help protect the soil structure because the soil isn't walked on and compacted.

HOW TO FILL A RAISED BED INEXPENSIVELY

Raised beds can be a pricey option. Here is how I fill a raised bed on a budget. Layer the bottom of your raised bed with cardboard, followed by a few layers of logs, branches, leaves, and other green and brown materials. I then fill the rest of the raised bed up to the top with a quality soil mixture. Over time, all of those logs, branches, and leaves will break down and add fertility back into your soil, and you aren't left spending a fortune on filling up the entire raised bed.

In-ground gardening

In-ground gardening is precisely what it sounds like. It is a method of gardening where you plant and grow directly in the ground, using the existing soil. Typically tilling is the method used to break up the ground initially. Once you've established rich soil, you can amend it with compost or another form of organic matter and use tools like a broad fork to work in that organic matter without disturbing the soil's ecosystem.

The pros and cons of in-ground gardening

Pros of in-ground gardening

- The costs associated with starting an in-ground garden are much lower than raised bed gardening.

- If you have nutrient-rich soil already established, all the hard work is done. You can continue improving the soil quality over time, but upfront labor and expenses related to soil building are not an issue. This makes in-ground gardening very appealing to a grower who wants to keep their budget low.

- This growing style provides you with more usable square footage for planting.

- It's not permanent. In-ground beds can easily be replaced with another crop or moved. This garden style usually appeals to the market gardener who wants to maximize yields and rotate crops.

- In-ground beds don't dry out as quickly as raised beds; therefore, you have fewer irrigation requirements.

Cons of in-ground gardening

- In-ground beds require an initial tillage. If you desire a no-till approach, in-ground gardening will not be the best option. At least not to start.

- Reduced drainage can be an issue. Depending on your region, soil type, and climate, in-ground beds are more prone to flooding.

- It will take longer to improve the soil quality, especially if you have less than ideal soil from the start. Continuously building and improving your soil will take time.

- Soil compaction can occur if beds are regularly walked on to cultivate or plant them.

- Due to the soil being at ground level, it will take longer to warm after a harsh winter; therefore, you will typically have to plant later than when planting in a raised bed that warms up faster in the spring.

In-ground growing options

As mentioned before, this is not a pass or fail, but more of a "good, better, best" situation.

Your region and growing zone will determine which crops are best grown in-ground vs. in raised beds. The crops listed in this sidebar are great for in-ground growing and will help you dig in and cater to your specific zone and your particular needs.

The plants listed below typically take up more space. It makes sense to grow them in an in-ground area, especially if you are starting a large number of plants.

Vegetables suited for in-ground gardening

Note: These varieties take up a lot of space and are best for in-ground gardening.

- Melons
- Peas
- Beans
- Corn
- Okra
- Potatoes
- Sweet Potatoes

More choices for in-ground growing

- Squash
- Zucchini
- Tomatoes
- Brassicas
- Flowers

All of these do exceptionally well when grown in-ground and given the correct soil conditions.

Something to remember: Almost any vegetable can be grown with the no-till approach either in-ground or in a raised bed. Reference your "why" to determine which option will meet your needs. That being said, some varieties do thrive in one style of garden over another. Plus, it depends on the variety. For example, larger melons take up a lot of space and are great for an in-ground garden, but some smaller-vined, personal-sized melons also work great on a trellis in a raised bed.

I use the Finesse broad fork on my farm weekly. Not only is it good for practicing no-till gardening, it is easy to use to flip and prep a bed for another round of crops.

No-till gardening

No-till gardening is not only a method I implement on my farm, but one I encourage any grower to consider as a viable option for their farm. No-till is an agricultural practice that does not disturb the soil through traditional tillage. Most raised bed gardens would be considered no-till because you aren't physically able to get a tractor into a raised bed and using a hand tiller would prove difficult and unnecessary.

I vividly remember going to my papaw's house every summer to pull weeds and work in his garden. He grew food using an in-ground growing method and a hand tiller. Growing up, every garden I visited was also designed for in-ground growing, and they used a tiller or tractor to turn the soil. This was normal to me. For many years, I didn't know you could have a garden without tilling. So naturally, when I decided to take the leap and create my first garden bed, I was on the lookout for someone with a tractor that could till up a plot of prime ground right in my backyard.

As you can imagine, not everyone was eager to load up their tractor and come to my house to make me a spot for my garden. I searched for weeks and weeks, and finally convinced a friend that I would trade him produce for his hard work. After all of the anticipation, I wondered...was this the only way? Surely, it didn't have to be this difficult to put in garden beds every season. I asked myself, what if I wanted to expand? Would I need to buy a tractor? Did I even have the budget to buy a tractor? All these questions lead me to the answer—no.

This was not, in fact, the only way to garden. I had the freedom to go against what I've always been told was the only way to do it, and I started to learn a different—perhaps even better—way. I learned about raised bed gardening and no-till gardening, and if I'm candid, this idea forever changed how I grow food.

The pros and cons of no-till gardening

Pros of no-till gardening

- By using a no-till gardening approach, you are increasing your soil's biological diversity and fertility.

- No-till doesn't disturb the beneficial microbes, insects, and mycorrhizae within your soil.

- The natural soil structure isn't compromised by tilling. The air and water pores within your soil stay intact, which reduces soil compaction and water runoff.

- No-till saves you money, labor, and time.

- Many tiny but mighty farms have higher efficiency and higher crop yields when using no-till methods.

- No-till gardening combats climate change because it reduces greenhouse gas emissions by reducing soil disturbances and keeps more carbon stored in the soil.

Cons of no-till gardening

- No-till gardening is a long-term return on investment. No-till mimics nature, which is beneficial. However, good things take time, and seeing the full benefits of your no-till practices will not come overnight.

- You may have increased use of chemicals if you are not growing organically. And if you are growing organically, you are perhaps going to spend more labor hours weeding and managing beds.

- Not all soils support this method of growing.

I recommend trying the no-till method whether you are new to gardening or an experienced gardener. No-till is slowly making its way into the gardening world as a real alternative—as it should be. No-till is no longer something that only market gardeners do; it's becoming an efficient and sustainable way to garden for many small-scale backyard growers.

No-till is no longer something that only market gardeners do; it's becoming an efficient and sustainable way to garden for many small-scale backyard growers.

We are only five chapters in, but I'm sure by now you know I like to do things outside of the box. When I started my first no-till garden bed, I had many questions, and people were confused by the concept. They wonder, is it in-ground gardening or something different? Now, I use this as an opportunity to share with others how it's different and how beneficial it can be for your soil and your farm.

Creating a no-till bed can be simpler than you think, see the sidebar for the steps required to create your own.

Building a no-till planting bed

Step one Mow or lay down a silage tarp on the area you want to turn into a no-till gardening space. I usually leave the tarp in place for 12 weeks. This will kill any existing weeds. If you have ground that has already been worked but just recently became filled with weeds, you can leave the tarp on for 2–4 weeks.

Step two Once the existing vegetation is dead, you can either lay down cardboard or paper that will act as a weed barrier. We use cardboard because we always have excess, and it can be sourced for free. Wet down the material of your choice. This helps it stay in place and speeds up the beginning of the breaking down process. The cardboard or paper will not only act as a shield for the weeds, but it is full of carbon-based organic matter that worms love.

Step three Next, choose quality topsoil mixed with compost, and begin piling that on top of your paper or cardboard. It will need to be at least 4–6 inches (10–15 cm) deep. You can top the soil with mulch, too, if you'd like. The bed is now ready for planting.

Each season you can "top dress" your beds by adding another layer of organic matter on top of the existing one. Wood chips act as a great mulch between no-till beds, and will break down over time, adding nutrients and organic matter back to your soil.

We practice no till on our farm. Here I am laying out a 24 x 50 silage tarp to naturally kill off the grass before planting. I secure the tarp with sand bags, logs from around the farm, and rocks.

When growing in containers, remember to water frequently to prevent the plants from drying out, and adding additional boosts of fertilizer.

Container growing

Remember when you had to have acres of land or a large backyard to grow food? Yeah, I remember that too. Thankfully, we've come a long way from that philosophy of gardening, and I know many of us are grateful for that.

There is an enormous need for people to be able to grow food in small backyard spaces, on patio balconies, and in the suburbs. Container gardening is a useful option well suited for this.

Container gardening certainly doesn't come without challenges, but it is a way to grow food in a small space for many people around the world. Although I live on 4.3 acres (1.7 ha), I still incorporate container gardening on my farm to add pops of color throughout the garden and to give my children opportunities to explore growing food in different ways. On any given day on our farm, you'll find grow bags lined up on the greenhouse deck and scattered throughout my garden. A large benefit of container gardening is the freedom to move your plants. If you live in a small space and lack optimum sun exposure in one single spot, you can easily move your containers to a different spot to ensure they are receiving adequate light. Being attentive to your plants and their needs is essential when growing in containers. Cater to those needs, and you'll see success.

Being attentive to your plants and their needs is essential when growing in containers. Cater to those needs, and you'll see success.

The pros and cons of container growing

Pros of container growing

- When growing in containers it's easy to relocate and move your entire garden. You can also bring the containers inside if your weather takes a turn for the worse.

- If you're gardening in an urban or small setting, containers allow you to max out the space. You can grow edibles even if there is no in-ground green space available.

- Containers are great for patios and balconies.

- Available in many sizes and designs, containers can add dimension and depth to your existing garden.

- Containers are much less expensive to start growing in compared to a bordered raised bed.

- You have control over the soil. If you notice the soil quality is low, you can efficiently work toward improving it or replace it with new soil at the start of each growing season.

Cons of container growing

- Containers dry out quickly. They can also heat up quickly in the summer—especially dark-colored or metal containers. Because of this, you will need to water more frequently, which can be time-consuming.

- When growing in containers, you will have to continuously fertilize; unlike raised beds and in-ground gardening, you aren't able to add mulch and amendments as easily, so fertilizers are needed instead.

- The soil used to fill your containers is critical. Don't use just any soil for your containers; use high-quality potting soil. Depending on where you live, finding good potting soil can be tricky. You will also have to replace the soil often as it will begin to settle over time or even dry out to the point where it needs to be replaced. Plus, your vegetable plants will deplete it of nutrients over time.

- Plant roots can quickly become confined in a container, causing them to become pot-bound and circle around inside the pot.

Terracotta pots are porous which allows good drainage for plants. The plus side to growing in these is that you can easily move them inside during the cooler months.

Understanding the varieties that will thrive indoors and in a short amount of time is crucial for success. Check out the tip box below for my favorite microgreen varieties.

Indoor growing overview

If you are just in the dreaming stage of your gardening or farming endeavors, this section is for you. Throughout this book, I will teach you how to grow abundant food in small spaces, but not everyone is ready to take that leap, and that is okay.

If you have a countertop, you can grow food. If you have room for a rack or shelf, you can grow food. If you have hands and a willing spirit, you, my friend, can and will produce food. Instead of being limited by your space, start thinking creatively. What can you do within your reach?

If your budget or space doesn't allow you to create an outdoor garden, don't be discouraged. Find another way. You'll be surprised by the amount of food you can grow indoors using grow lights. Indoor growing, for many, is a great way to garden.

Where to start

In the transition season on our farm, before the garden was up and going, I started to grow and sell microgreens. I was very encouraged by the amount of food I could produce within my home in a small space. We found many uses for microgreens—in our smoothies and salads, mixed in omelets, and we ate them raw for a snack. Even if you cannot grow outside at the scale you want, this is an excellent option to encourage you to grow indoors.

Many online gardening stores sell indoor microgreen-growing kits with everything you need to get started. I've grown microgreens of various herbs (basil is a must so that I can make homemade pesto), baby greens, and lettuce varieties.

Yes, you will be limited on what you can produce if you grow indoors, but it doesn't mean you have to be out of the gardening game altogether. Sure, you won't be growing tomatoes and peppers or even beans on a trellis *in* your home, but plenty of plants will thrive.

Greens do great indoors and allow you to explore different ways to eat and prepare them. When I was exploring growing food indoors, before I even had an outdoor garden, I took the time and researched it, so when that day came to start an outdoor garden, I would feel equipped.

I bought books upon books, listened to podcast episode after podcast episode, visited the farmer's market every Saturday, and asked gardeners for their tips and advice.

Instead of feeling limited, I felt motivated and even inspired. If I could grow food in my house and sell it to my community, I could only imagine what I would be able to do when I had actual ground to grow food in. Hold strong, gardener. You are capable of mighty things.

My favorite microgreen varieties

Spicy Mix: arugula, mustards, and radish

Smoothie Mix: wheatgrass, pea shoots, and cabbage

Salad Mix: beets, red cabbage, and broccoli

There are endless possibilities for different microgreen mixes. I always mixed my own seed blends based on my family's preferences.

More edibles to grow indoors

You don't have to feel limited by a lack of outdoor growing space. You can grow plenty of food right on your kitchen counter; here are a few more things that will thrive inside.

- salad mixes
- spinach
- radish
- carrots
- scallions
- micro tomatoes

While these will take a bit more love and attention than microgreens, it is possible. The only thing keeping the seed from sprouting is your unwillingness to plant it. So, here's my permission to plant the seed and watch it grow.

Which gardening style is best for you?

After reviewing each of these gardening styles and what they offer you and your small farm, you are now tasked with the joy of figuring out which one (or ones) you want to implement. Perhaps you may even want to practice all of them on your small farm. That's okay! You can start off choosing one method and change to another process down the line. There is no pass or fail here, only learning and growing as a gardener.

> "If you look the right way, you will see the whole world is a garden."
> —Frances Hodgson Burnett

As mentioned earlier in this chapter, some gardening styles offer more efficient production, while others offer an aesthetic appeal. I encourage you to try each of these growing practices in one form or another to get a basic idea of which one will work best for what you are trying to achieve. Regardless of which of these tactical approaches you take, I want to encourage you to fall in love with what you are doing in the garden.

All of these styles—raised beds, in-ground gardens, indoor growing, and container gardening—can benefit a small farm in many ways, but the love you cultivate by growing will be what draws you to come back to the space on the hard days.

Start herbs indoors, build raised beds, and prepare some in-ground garden space to see which one resonates with you. Hold onto the knowledge that you can change your mind again and again.

If you are reading this and are in the dreaming stage of your small farm, I can tell you, this is my favorite part of the process. When you create your first garden beds, you will remember them for a lifetime. These beds will be the ones that turned you into a gardener, teacher, and listener of this Earth.

Producing food is exhilarating, no matter how it is grown, especially when it is shared with family and friends and fresh from your backyard. Starting and expanding your growing space is a joyous ride. It is worth all the sweat and tears you'll shed in the process.

Now it's time to roll your sleeves up, get your hands dirty, and work. It will be challenging, yes, but the reward is worth it. Strap on your boots and get busy, friends. You've got a story to tell about a bountiful garden in your near future!

I look forward to seed-starting every year. Make sure you know when your estimated last frost date is, and check your calendar to make sure you are starting your seeds at the appropriate time.

CHAPTER SIX

Growing From Seed

We have reached my favorite chapter in the book. I hope it is as fun for you all to read as it's been for me to write. Who doesn't love spending their day covered in dirt in the greenhouse? If that's not you, hold tight, I might change your mind by the end of this chapter.

This chapter covers topics from the benefits of starting seeds to step-by-step how-to's, deep-diving into the differences between hybrids and heirloom varieties, and so much more. I think back to when I started my first garden; I had a handful of watermelon seeds and some plants from the local hardware store about 5 miles (8 km) up the road. I never purchased organic plants, I treated them with harsh chemicals, and my garden sure as heck didn't thrive like I thought it would. While it did grow wild and free, it left a lot to be desired.

A few years later, I transitioned my entire life to a more sustainable, organic, and healthy lifestyle. At that time, I started leaning into growing and expanding my garden, but I also spent time and diligence researching and understanding heirloom and hybrid varieties and which of them would be better for my tiny organic farm.

I could spend an entire chapter explaining all the cool facts about seed starting, but to spare you what a "fun" afternoon with Jill looks like, we'll keep to the essentials. I urge you to look up more fun and interesting facts about the seeds you plant, such as where they are bred and sourced, how they've evolved through the years, the various sizes of seeds, information about rare varieties, and the endless growing possibilities available to you.

Oh, so, back to those seed-starting essentials I promised to stick to. Let's start with why we should start our own seeds instead of just buying plants. Isn't that too much work? That's certainly what I thought many moons ago. Now I can't imagine a season passing by without a greenhouse full of freshly seeded trays. Here's why I choose to start from seed and hope you will do the same.

If you've never grown hybrids varieties before, here is your permission to buy the seeds and see how they will forever change how you view production crops.

Why grow from seed

Starting seeds is an excellent way to get a head start on your growing season. Plus, it's more cost-effective to start your plants, especially if your goal is high production. One potted plant from a nursery will likely cost you more than an entire package of dozens or even hundreds of seeds. One plant versus 50–200 seeds? That's a no-brainer for me. The good thing, too, is that planting from seed gives you the liberty to fail and try again. The cost of starting seeds (even for the second or third time) is significantly less than buying multiple potted plants from a nursery.

My favorite thing about starting seeds is the thousands of varieties I can choose from; these are unique varieties I would not be able to find as plants at my local nursery. Tromboncino squash has made its way into my garden for the last four years, and I can assure you I would not be able to find a Tromboncino squash plant for sale in nurseries anywhere near me.

I grow food for more than the yield. I grow food for the experience. The experience of trying new varieties, of seeing vegetables I'd never know

if I didn't grow them for myself. That feeling of growing something new that I started from seed for the first time is nothing short of extraordinary. Starting seeds does not have to be a terrifying endeavor, but here are a few things to consider before jumping in.

Growing from seed allows you to get a head start on the growing season, to grow more unusual varieties, and to get more plants for your money.

I start my seedlings in a range of trays and various soil block sizes. These 2.5 inch (6.5 cm) pots are excellent for things that I know will be in containers longer, such as these brassicas.

Be sure to clearly label all pots when
starting seeds. If you're using soil
blocks, you can label the entire flat.
Trust me, this practice saves a lot of
confusion down the line.

Starting seeds indoors

Starting seeds indoors has many advantages, especially if you aim to grow intensively and implement succession planting. You can either start your seeds in a greenhouse (which I'll discuss later) or indoors, using racks and grow lights. Reference the equipment sidebar below to set yourself up for success.

Equipment needed for starting seeds indoors

- Containers (your choice of size) with drainage holes

- Deep 10 x 20-inch (25 x 51 cm) plastic trays (with no holes for bottom watering and housing your seedlings if starting them in single pots)

- Seed-starting potting soil of your choice

- Metal racks (if growing a large quantity) or shelves

- Grow lights (these can be anything from inexpensive shop lights to fancy LED lights specific for growing plants)

- Seedling heat mat (optional)

- Humidity dome

- Labels

- Markers

- Seeds

There is no right or wrong way to start your seeds. Some of mine are started in the glass greenhouse in the cottage garden. The majority are started in my 14 x 36 feet (4 x 11 m) seed-starting tunnel, but every season I ambitiously start a few indoors using a rack and a grow light.

Get a head start

By starting your own seeds, you are able to get a head start on the season. While it still may be dark and cold outside, you can be cranking out seedlings in your greenhouse or indoors under lights. This is a huge advantage because your seedlings will be mature and sturdy by the time you are ready to move them out into your garden.

Cool-season crops like brassicas take months from seed sowing to harvest. If you were to plant seeds of these crops outside after you are clear of your last spring frost, they will likely not reach maturity before it gets too hot and they will stop producing. By starting these varieties ahead of time indoors, you ensure a yield.

Heat-loving plants such as tomatoes and peppers greatly benefit from getting a head start indoors as well. Plus, who doesn't love spending those cold, dark winter days in a greenhouse starting seeds? It is one thing I look forward to every winter. It reminds me that the promise of spring is quickly approaching.

Control the elements

I live in Central Arkansas, where the weather can be, well, quite unpredictable. Just this past winter, we went from a high of 80°F (27°C) down to a low of 17°F (-8°C) overnight. My poor, overwintered flowers had no idea what hit them.

The ability to control the environment and climate where you are growing your seedlings is crucial. You can utilize heat mats, quality potting soils, grow lights, humidity domes, etc. to help control the seed-starting environment. When you start your seeds indoors, you have the ability to control the elements. You may notice faster germination than when sowing outdoors because you can regulate and adjust heat, light, and humidity levels.

Most cabbages take an average of 70 to 100 days to go from seed to harvest. It's important to understand this when you are starting seeds so you can calculate ahead when you can expect a harvest.

When starting seeds in a greenhouse, grow lights aren't necessary, but heat mats can be beneficial for heat-loving crops like tomatoes and peppers.

When you start your seeds indoors, you have the ability to control the elements.

The key to successfully starting seeds indoors is to provide the right temperature and high humidity. Germination chambers are a tool you can use to give both of these elements to your little plant babies as they wait to emerge from the soil.

What is a germination chamber?

A germination chamber is an enclosed compartment used to house your seedlings. It provides optimum humidity and temperature control. By giving the opportunity to regulate the temperature and humidity of your seedling, they experience excellent germination rates and

growth. You will only need to keep the trays in the germination chamber until they have germinated, then they will need to be moved to the light.

You can find germination chambers online at most garden supply stores, and some local nurseries carry them as well. Usually, you have the best selection when purchasing online. Germination chambers range in size based on how many trays you want to fit in them. The standard has 4–5 shelves.

Protect your seedlings

Baby seedlings can be vulnerable to the elements if planted outdoors. I speak from experience. There's no greater garden heartbreak than sending your little plant babies outside only to find after a few days that they've been eaten to a nub by cutworms or other pests. There are some natural ways to combat cutworms by placing diatomaceous earth around the base of the plant. You can also handpick them when you see them. And Bt (*Bacillus thuringiensis*) is an excellent insecticide spray for cutworms (and other destructive caterpillars). The key to success with any of these is consistency. Treat or handpick often.

Growing your seedlings indoors or in a greenhouse allows them to grow strong and stay protected from harsh weather and pests during their most vulnerable time. When your plants have reached the time for transplanting outdoors, they will be able to combat pests on their own, with the occasional handpicking or treating, because they are not nearly as susceptible as when they are freshly sprouted.

Sprinkle dipel dust or diatomaceous earth on seedlings early in the morning or late in the evening to combat caterpillars.

Grow with the seasons

My favorite time to garden isn't summer. It's in the fall and winter. Please don't hear me wrong. I love the summer. Waking up early, walking through the garden with coffee in hand, and watching the sunrise is a sight you never get tired of. And after a long, hot day of harvesting, lying in the garden, looking up at the star-filled summer sky feels like something everyone should get to experience in their lifetime. It's bliss—hot, exhilarating joy. But then summer leaves, and you have the promise of fall, a time to slow down and enjoy the garden in a way you simply cannot in the summer. It is a time to reflect and a welcome shift.

We want to utilize all the seasons on our farm, meaning that when one crop is finished producing, we replace it with another. Starting our own seeds allows us to let our summer crops continue to grow to their full potential while we nurture the fall and winter seedlings in the greenhouse. We've spent a lot of this chapter expressing the benefits of starting your seeds either indoors or with the use of a greenhouse, but if you do not have a greenhouse or space in your home to start seeds, there is another way.

Direct seeding

Let's discuss what it means to direct seed. Simply put, it is a method where you plant the seeds outside, directly into the ground. Reference each seed packet for spacing and planting instructions. With this method, seeds will germinate, sprout, and produce food. Then the plant dies, and you start the process again. Sounds pretty easy, right? It is. While you don't get a head start on your growing season when direct seeding, you do save time and eliminate steps like hardening off (see the hardening off section later in this chapter) when you direct seed your crops. Some plants do not like being transplanted, so for those varieties, direct seeding is always your best option for a successful crop.

Direct seeding will also help keep your systems efficient (I'll explain how in chapter 7). With good timing and techniques, you should have no issues with germination and there is minimal cultivation needed to direct seed.

One thing to note would be how you plan on seeding your beds. You could do this by hand or with a mechanical seeder. If you are directly seeding several beds, I would think about the time required to do it by hand and perhaps consider investing in a mechanical precision seeder.

Here is a list of a few crops that benefit from direct seeding:

- Carrots
- Radish
- Rutabaga
- Turnips
- Garlic
- Potatoes
- Arugula
- Parsnips

Most of these are root vegetables and tubers that typically do not like to be transplanted, and if you try to, often the plant will be stunted or malformed. It might even die.

Sometimes I like to live life on the wild side. When someone tells me "You can't do that" or "Don't plant that," the inner rebellious child springs forth, and I want to do just that—test the waters, figure out for myself why something should not be done, and then, have an assessment of my own.

The fall and winter gardens are less demanding than those of the spring and summer. The shift is always welcomed.

When I read "best direct-sown" on a seed packet, that's when that inner rebellious child comes out, so if you are like me and prefer to test the waters, here are a few crops that I have grown indoors and then transplanted out into the garden, despite the explicit instructions on the seed packet to direct seed. I have found great success doing this for some vegetables.

Crops to experiment with starting indoors before transplanting them out into the garden:

- Beets
- Spinach
- Peas
- Kohlrabi
- Sunflowers

Cattle panels are an excellent resource to grow food vertically. There are many options for what will grow on them. My favorites are vining squash and green beans. When you grow vertically, it adds an ease to maintaining your plants and makes harvesting a breeze.

Tiny tip

Plant your seeds twice as deep as the size of the seed itself.

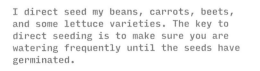

I direct seed my beans, carrots, beets, and some lettuce varieties. The key to direct seeding is to make sure you are watering frequently until the seeds have germinated.

Successful direct seeding

Direct seeding is a simple, straightforward option, but to ensure success, below are a few tips.

- If your ground is hard and dry, water evenly before planting. Keep the soil moist before seeding to help speed up germination.

- After seeding, it is also essential to maintain moisture in your soil. This requires you to check your beds often, but it is a step you do not want to miss for crops that are slow to germinate.

- If you have irrigation set up in your beds, you can set it up on a timer; that way, you can water at the same time each day for the same length of time. However, I typically hand-water for newly planted beds to ensure everything is good and moist when it needs it.

- Even if you have irrigation in place, it is important to regularly walk through your garden and lay eyes on everything. If you notice some spots are dry, grab a hose. If some beds are too wet, this would be the time to adjust your irrigation timers. Keep an eye on the state of your farm. Visual inspections allow you to adjust things as needed and ultimately it promotes success in the garden.

Space matters

Knowing and optimizing your space is crucial when it comes to direct seeding. You could plant densely, packing your beds with as many seeds as possible, but then you would struggle with being able to work around your plants. And, spacing that is too tight could also lead to reduced airflow and therefore increase potential pest and disease issues. On the flip side, you could space your plants too far apart and then spend unnecessary time weeding the open ground that's been left unplanted. As you can see, this is a delicate dance you will have to learn to balance. Proper spacing is the best fertilizer.

Ginger is actually not a root: it is technically known as a rhizome and can be easily grown from ginger bought at the store.

Proper spacing is the best fertilizer.

This balance took me a while to find, and that may be the case for you, too. Plant seeds close enough to help prevent weeds but allow them enough room to fully mature. Don't forget, if you are using tools for cultivation, space your rows wide enough to use them. The goal is to make your job easier, not harder.

Pros & cons of indoor seed starting and direct seeding

There are certainly many pros and a few cons to both of the seed-starting methods mentioned above. Crop varieties and the timing of your planting play a prominent role in deciding which approach, if not both, is best for you and your garden.

Kohlrabi is known as a German turnip.
You can consume it in many ways, and it
can be eaten raw or cooked. We prefer to
ferment it. It also does well as a potato
substitute. This fun plant is a biennial
and is best grown as a cool-season crop.

How we track seed sowing

These two charts are examples of how we plan and track the sowing dates for the crops we grow on our farm. It helps us manage the timing and encourages a long harvest period. The first chart tracks our indoor seed sowing. The second chart is how we track the seeds we direct sow out into the garden. You can use our charts as a starting point for developing your own tracking system. We organize by the sow date, but perhaps you would prefer to organize by the variety or even by the type of cell tray you use. There are many options. It may take a while to develop your own system, but it's worth the effort.

Sample Indoor Seed Starting Tracking Chart

Vegetable	Variety name	50 cell tray	128 cell tray	264 cell tray	Seeding date
Onion	'Zoey'			8	January 6
Onion	'Cabernet'			8	January 6
Parsley	'Giant of Italy'		5		January 13
Lettuce, Head	'Mirlo'		12		February 10
Lettuce, Mix	Salanova® Foundation Collection			12	February 10
Broccoli	'Belstar'	12			February 17
Tomato, Cherry	'Esterina'	2			February 24
Tomato, Cherry	'Purple Bumble Bee'	2			February 24
Tomato, Cherry	'Sungold'	2			February 24
Tomato, Heirloom	Heritage Mix	1			February 26
Tomato, Heirloom	Heirloom mix	1			February 26
Pepper, Tricolor	'Carmen'	4			March 3
Pepper, Tricolor	'Escamillo'	4			March 3
Pepper, Tricolor	'Milena'	4			March 3
Cilantro	'Santo'		4		March 3

Vegetable	Variety name	50 cell tray	128 cell tray	264 cell tray	Seeding date
Eggplant	'Traviata'	5			March 10
Cucumber	'Excelsior'	2			March 31
Cucumber	'Socrates'	2			March 31
Squash, Yellow Summer	'Tempest'	3			March 31
Squash, Zucchini	'Dunja'	3			March 31
Squash, Acorn Winter	'Tuffy'	6			April 28
Squash, Butternut Winter	'Waldo'	6			April 28
Lettuce, Head	'Red Cross'		12		May 10

Sample Direct Seeding Tracking Chart

Vegetable	Variety name	Seed quantity	Direct sow date
Arugula		2 packets	February 21
Arugula	'Astro'	3 packets	February 21
Radish	'Sora'	5 packets	February 21
Spinach	'Corvair'	3 packets	February 21
Carrot	'Hercules' (Main Crop)	2 packets	March 3
Carrot	'Napoli'	3 packets	March 3
Beet	'Babybeat'	2 packets	March 21
Beet	'Chioggia'	3 packets	March 21
Beet	'Red Ace'	4 packets	March 21
Green Bean	'Provider'	6 packets	April 28
Green Bean, Pole	'Musica'	5 packets	April 28
Squash, Zucchini	'Green Machine'	25 seeds	April 28
Squash, Pattypan	'Sunburst'	25 seeds	April 28

Setting your beds up on drip irrigation makes a significant impact on the growth of your plants. For the backyard hobbyist juggling jobs, etc., not having to worry about watering your plants will set you up for success early on.

After you evaluate whether you will be starting your seeds indoors in a greenhouse or your home or by direct seeding, you can start to pivot and spend time flipping the pages of seed catalogs and deciding which varieties you want to grow. As you read the descriptions in the seed catalogs, you will see things like "days to maturity," "spacing recommendations," and "planting practices." These are good to take note of when deciding what to grow. Knowing your frost dates will also help you decide if you have the time and climate to grow each of your desired crops.

Frost dates

After being cooped up in the house all winter, it's easy to go wild sowing seeds but hold your horses. Nothing is worse than starting seeds too early, nurturing them to transplant size, and then putting them outside only to watch them get nipped by ol' Jack Frost.

Knowing your average first and last frost dates is essential for your seed starting endeavors. If you do not already know when your average last spring frost date and first fall frost date is, you can do a quick internet search to find the answer. Back in my wild days, I always took my chances and planted early. It didn't take long for me to learn from my mistakes. These dates can seem negotiable, especially if you live in a warmer climate, but you very much need to take them seriously or you may end up with a sad start to your season.

Once you have figured out your average last and first frost dates, you can finally embark on your seed starting journey. Check the back of each seed packet for information on when to plant each variety based on your specific last frost dates. For example, the packet may say "sow seeds indoors 8 to 10 weeks before your last expected spring frost" or "directly sow seeds out into the garden after the danger of frost has passed"—which means plant them after your last frost date has passed.

A long-handled stirrup hoe is my go-to. The stirrup-shaped head allows you to cut the root system off the weed without interfering with other nearby production crops. It is lightweight and easy to use.

You can also determine the best planting time based on the average first fall frost date. I like to start by figuring out how many days until the variety matures (noted as "days to maturity" on the seed packet). Let's say you grow a variety that takes 60 days to mature. Turn your calendar to the date of your first expected fall frost and count back 60 days. Now you know when to start those seeds. I also like to keep a gardening journal/planner and jot down all the information about when I start each type of seed for all of the various varieties I grow.

Once you have established the seeds you want to grow and the proper timing for planting them, here comes the next steps.

Choosing containers for indoor seed starting

The options are almost endless for containers to start growing your seeds. Traditional plastic nursery trays are a great choice. They can have four-celled or six-celled insert packs and contain from 12, 36, 48, or 72 cells per tray, or they can have smaller "plug" cells with 128–512 cells per tray. These types of nursery plastic trays and inserts are what I typically use for starting my seeds. You could also use recycled plastic cups, compostable pots, egg cartons, etc. The list of possibilities goes on and on. If you are using a recycled pot, sanitize it properly first to prevent any diseases tagging along.

In truth, recycled cups or egg cartons aren't the best option for the intensive home gardener. Cups take up a lot of space when starting plants indoors or in your greenhouse. I have recently used pulp pots and found that they don't dry out as quickly but certainly come with their own list of obstacles as well. I have found that the pulp pots don't hold together well. For instance, we used these for a plant sale and after we watered them well in the morning many of the pockets would tear apart when someone went to pick them up. Although they don't dry out as quickly as the peat pots, they still dry out faster than if we were using a plastic container.

Remember these are suggestions, not rules you have to abide by. Try growing new crops in different seasons and take note of how they performed.

So even though there are plenty of options, you will need to try a few out to see what suits your needs the best. It is important to note that if you are using cartons or recycled cups, you will need to poke holes in the bottom to ensure proper drainage.

Soil blocks

Have a seed-starting notebook or create a new note on your phone to easily keep a record of when you started your seeds.

Another option I have been experimenting with is soil blockers. For the home gardener growing many starts, I highly suggest looking into them. Instead of using plastic pots or nursery trays, you create blocks made out of growing medium and water. My initial reason for venturing into soil blocking was my lack of space. When we began our first growing season on our new farm, I had a seed-starting tunnel that was still in the process of being built, and I had to think outside of the box for ways to save space. A friend introduced me to soil blocking, and it shifted how I thought about seed starting. I was floored at how many seeds I could start and how little space they took up.

Now, I have four types of soil blockers, and they each serve a different purpose. I have the single soil blocker, the hand-held 4 soil blocker, the stand-up 12 soil blocker, and the hand-held 20 mini-soil blocker.

I start most of my flowers, spinach, and lettuce varieties with the hand-held 20 mini-soil blocker. The hand-held 4 soil blocker, which is bigger, is used to start things like my tomatoes, peppers, squash, and zucchini.

The single soil block maker measures 4 x 4 x 4 inches (10 x 10 x 10 cm), which is what I transition my pepper and tomato seedlings into once they become a bit stronger and bigger and outgrow the smaller blocks.

The benefits of using soil blocks extend beyond space saving. They also reduce plastic usage. Plus, your transplants are often healthier and more vigorous due to their oxygenated roots. You also reduce transplant shock significantly when using this method.

I've been starting my own seeds for years now, and I have been pleasantly surprised at how healthy the transplants look in the greenhouse, and how well they transition into the field.

If you are trying to start seeds on a budget, you can reuse old seed containers or visit your local nursery and ask if they have any extra trays to get rid of. You would be surprised at the amount of trays you can snag for free.

top: Here I have the handheld 4 blocks up-potted into the single soil blocks. These work well for crops you want to mature more before transplanting. Peppers and tomatoes are the only crops we are up-potting into the single soil block.

left: The 20 mini block takes up the least amount of space but requires the most attention due to the seedlings drying out faster.

Whether you are starting your seeds in containers or soil blocks, knowing the adequate size is necessary.

The right container size matters

Choosing the right size container is essential. If you decide to start your seeds in a small container, you will likely need to "pot up" your seedlings and move them to a larger pot. Otherwise they will become root-bound before you can transplant them out into your garden. If you choose a container that is too large, you will be wasting soil, and the containers will take up more space than necessary. If space is something you have a limited amount of, think through these considerations.

This is where the soil blocks really start to shine. Start with the smallest block and continue to pot up as the seedlings grow—maximizing your use of available space and using just the right amount of soil.

Soil for success

Which is better for growing seeds—seed-starting mix or potting mix? That is a great question. I have started in both mediums, and it does make a difference. Seed-starting mix will be labeled as such and usually is a bit pricier than potting mix, but it will be free of larger pieces and be light and fluffy.

The handheld 4 is used to up-pot your 20 mini seedlings. This soil blocker leaves an indention in the new block the same size as the mini 20 to make up-potting faster and more efficient.

In contrast, potting mix often contains bigger particles, and you will need to sift these out before starting your seedlings because they may cause root distortion. Read the label and ensure that the soil medium is free of any synthetic fertilizers that could stunt or burn young plants. A seed-starting mix doesn't usually contain fertilizer, so if you plan to have your seedlings in this mix for

an extended period of time and you start noticing the plants showing signs of nutrient deficiencies (smaller plants, pale leaves), feed them with a good fertilizer. My go-to is fish emulsion. Most potting soils do contain starter fertilizers so adding additional fertilizer to these mixes may not be necessary.

The single soil block is the beast of all soil blocks. When our transplants are ready to head to the high tunnel they are thriving, never root-bound, and hold up extremely well.

Seed-Starting Issues

Issue	Symptoms	Treatments
Nitrogen deficiency	Discolored leaves, usually yellow, indicate a nitrogen deficiency.	Treat nitrogen deficiency by starting a fertilizing routine as soon as the seedling's first true leaves have developed. Follow fertilizer label instructions and remember—less is more, especially for young seedlings.
Phosphorus deficiency	Purple discoloration on the leaves. Leaf tips may turn brown and eventually die. Most commonly, leaves turn dark green and appear stunted.	Add bone or blood meal, rock phosphate, manure, or a phosphate fertilizer to the soil. Apply the fertilizer to the root zone (at the base of each plant) and water in well.
Leggy plants	A lack of light causes leggy seedlings when the plant tries to accelerate its height to get closer to the light source. Tall, thin, and floppy seedlings are signs that you have leggy plants.	Make sure your seedlings have sufficient light. If growing indoors, put your seedlings in a south-facing window, and if using a grow light, it should remain just a few inches above the top of the seedlings. Adjustable lights are ideal so you can move them up as your seedlings grow.

Poor germination due to seed age or health	Poor germination is merely not having your seeds sprout. You may discover this if your seeds are "too old" or they have not been stored properly. I encourage you to check the "packed for" year on the back of the seed packets before purchasing and select seeds from a company you trust.	Pre-germinate a small number of seeds in a damp paper towel placed in a plastic bag to check viability if you are unsure whether your seeds are too old. Some seeds last for a long time, so very rarely do they not germinate. But others have a shorter shelf life.
Poor germination due to improper planting	Seeds contain enough energy for the seedling to sprout and grow to the soil's surface, where they will begin photosynthesizing to create more energy. If your seed is planted too deeply, it may die before it breaks through the soil surface.	Seeds should be planted twice as deep as they are wide. Some seeds require light to germinate, while others require exposure to a cold period (this is known as stratification). Ensure each seed's germination requirements are met. Careful watering and bottom watering also ensure seeds are not pushed deeper into the soil during watering.
Poor germination due to incorrect temperatures	Less than ideal temperatures will slow germination. Cool temperatures combined with wet soil can cause seeds to rot if exposed to these conditions for a long period of time. However, in most cases, the seeds will just wait for the right temperatures to start growing.	Use seedling heat mats (more on these on page 132) to raise the soil temperature 10°F (5°C) above room temperature to speed germination. Remove the heat mats once the seeds have sprouted.
White or green mold on the soil	Mold on the soil surface typically means that your soil is too wet. Mold is caused by a fungus that needs moist environments to thrive.	This is typically an easy fix. Use a fork or tool to scratch the soil surface to increase aeration. Add a fan or open a nearby window if you're growing in a greenhouse to encourage proper airflow. I suggest bottom-watering your seedlings and doing so only when you start noticing your soil feels dry. Avoid overly wet soil.
Damping off	Damping off is a soil-borne fungal disease that affects seeds and new seedlings. The disease rots the stem and root tissues at and below the soil surface. In most cases, infected plants will germinate and come up fine, but within a few days, they become water-soaked and mushy, fall over at the base, and die.	There is no cure for seedlings that have already suffered from damping off. However, to prevent it from happening to your seedlings in the future, provide sufficient air circulation, either from a fan or by propping up the lid on your germination tray. When starting seeds, use good quality soil that allows proper drainage. Sow seeds thinly to prevent overcrowding which can lead to humid, moist conditions.
Fungus gnats	Fungus gnats are tiny mosquito-like insects, about one-eighth of an inch (3 mm) in length, that feed on your potting soil. They are hard to see flying around, but you might notice them on your newly sprouted seed trays.	Make the habitat inhospitable for fungus gnats to live. This means allowing the soil to dry out. Fungus gnat larvae live in the soil and need moisture. If you let the soil dry out completely before your next watering, it will get rid of the larval gnats and discourage more adults from laying their eggs in the soil. You can also mix a cup (236 ml) of water with two or three drops of liquid dish soap. Spray the top of the soil with this solution to kill the larvae. Repeat this process again in a few days. Alternatively, scrape off and dispose of the top layer of soil outdoors. The eggs can still hatch, but by moving the soil outdoors and away from your seedlings, they will not cause issues with your newly sprouted seedlings.

Continuously up-potting is another step in your seed-starting journey, but in my experience the seedlings are stronger, require fewer nutrients because they are continuously being up-potted into new soil, and experience less transplant shock.

Optimizing growing conditions for indoor seed starting

Once you've seeded your trays, what then? There are a few things to keep in mind to ensure you are providing your seedlings with optimum conditions to germinate and thrive. Let's discuss these things.

Heat is necessary

For your seeds to germinate, they will need to be kept warm and moist. The cooler your soil temperature is, the longer you will wait to see seedlings sprout. If you are anything like me and you stalk your greenhouse to spot the first seedling to pop out of the soil, I don't want to wait any longer than I have to. Seedling heat mats will speed up germination and are a great tool to keep soil

If you discover large chunks of debris in your soil medium, use a mesh sifter to separate out the larger chunks that aren't ideal for seed starting.

warm without overheating and drying it out. These can be found in stores or online at most places that sell gardening supplies. They are relatively cheap and last a long time.

Tiny tip

When starting your seeds, you may come across some flowers or perennial plants that require **cold stratification**. This was a foreign concept I knew nothing about until I started a few particular seeds several times and had zero germination. After some investigation, I realized I was missing a very crucial step in their growing process. The seeds of many long-season perennials and biennials require a cold stratification process (exposure to prolonged cold temperatures) in order to break dormancy. Also the seeds of cold-season annual edibles like spinach, arugula, and onions greatly benefit from exposure to a cold period, too. You can store all your seeds in a freezer to ensure they have gone through this process, or purchase seeds that have been pre-stratified (as per the seed catalog).

Light requirements

When I started seeds for the first time, a greenhouse was something I had only dreamt of. I wasn't aware that I could start seeds without one. So when I did receive this information from a dear friend, my eyes lit up. Do you mean I didn't have to invest hundreds or thousands of dollars into erecting a greenhouse to start seeds?! Nope, a greenhouse is definitely not needed to start seeds. You can start seeds without artificial light if you have a sunroom or a south-facing window with good light. Or, if you don't have that available, you can purchase a new or used wire rack, and add some grow lights or shop lights and be in business for under $100. Most seeds don't actually need light for germination itself, but it is vital for the next step of the process, which is photosynthesis and your plant's growth.

If using grow lights, hang them from the racks above your seedlings so they are suspended around 2–3 inches (5–8 cm) above the plants. We kept ours on a pulley system, so I would raise the lights as the plants grew. If you do not do this, your plants could become "leggy," which means they are reaching for more light (see chart on page 130). This causes their stems to be extremely thin and fragile, resulting in plant breakage.

Constantly check your seedlings for deficiency. Each season I struggle with a phosphorus deficiency in my tomatoes and the underside of their leaves start to turn purple. Within the first 24 hours of adding a boost of fertilizer I notice the issue starting to resolve.

The key to fertilizer is to not overdo it. A little goes a long way; treat only when you see a deficiency and always read the back of the label.

Essential equipment

Despite what you might think, you don't need a lot of equipment to start seeds, and most of the items will be a preference, not a necessity. Knowing how you intend to start your seeds determines the equipment you'll need. For example, if you're only starting in a soil block, you will not need to source containers. You won't need racks or grow lights if you're growing in a greenhouse.

> **Despite what you might think, you don't need a lot of equipment to start seeds, and most of the items will be a preference, not a necessity.**

However, items such as seeds, labels, and a water hose with a fine nozzle are essential. Try not to overthink what you *actually* need, though. For labeling, you can use a piece of tape and a sharpie, a cut-up recycled yogurt container, or even popsicle sticks. It doesn't need to be fancy to be efficient.

Heat mats came in various sizes. I prefer the 6-foot (2-m) roll because it fits the shelf of my glass greenhouse perfectly. If you are only starting one or two trays, the 2-foot (60-cm) heat mats would be best.

How to start seeds

Once you've collected the materials to start your seeds and established how you will be growing them (indoors or in a greenhouse) you can begin the fun part, getting your hands dirty.

Seed starting is a great way to reuse old items and repurpose them. These tags were made from yogurt containers. Remember the 3 R's (reduce, reuse, recycle) to eliminate waste on your farm.

Step-by-Step Guide to Seed Starting

Step 1 Moisten your soil until it's damp, but not saturated.

Step 2 Fill your trays or pots with the soil of your choice, tamp lightly making sure there aren't any big air pockets.

Step 3 Label each tray or pot with the variety and planting date.

Step 4 Make a hole in each cell, using your finger or dibbler.

Step 5 Plant the seeds according to your seed packet.

Step 6 Lightly cover the seed with more soil, unless the seed packet has instructed otherwise.

Step 7 Water from the top or place trays or pots in the bottom-water tray to absorb water.

Step 8 Move to a heat mat or warm location in your home.

Step 9 Once seeds have germinated, move them under grow lights or to a sunny greenhouse and remove the heat mat.

Step 10 When soil becomes dry, water gently and evenly.

Step 11 When young seedlings begin to outgrow their pots or tray, or when you are free of your last frost, being the hardening-off process.

Step 12 Transplant out in your garden.

If you have planted more than one seed per cell container, you will need to "thin" them out. You can pinch one off at the soil surface, keeping the root system intact. Or, you can gently separate the plants and pot up the excess.

Thinning

The need to thin your seedlings only happens if you plant more than one seed per cell or container. Suppose I had seeds for an exceptionally long time and I was worried about low germination rates. In that case, I might sow more than one seed per container. Or, sometimes the seeds are so tiny that more than one gets seeded in each container by accident, which is okay. You can choose to thin by cutting the secondary, weaker plant off at soil level (do not pull it as that disturbs the root system of the remaining plant). The other option would be to pop the entire group of seedlings out of the container and separate them. Then repot each seedling into its own container. It is best to either thin or separate them because the plants will compete for nutrients, light, and resources.

Sometimes gardeners grow multiple seeds in a 5 x 5 inch (13 x 13 cm) cell growing tray or a 10 x 20-inch (25 x 51 cm) plastic flat and then separate the seedlings as they start to get bigger. I have done this in the past with flowers, and it worked pretty well for things that I could transplant out into the garden without needing to pot on. It saved soil, containers, and time.

Hardening off your plants

Do I need to harden off my plants? That is a question I've answered many times in my years of growing a garden, and the short answer is yes. When you think about your seedlings all cozy in your greenhouse or tucked snug under your grow lights for weeks and weeks, throwing them outside, unexposed to the elements, would subject them to a real shock. Most times, if your seedling has experienced severe shock, there is no recovery, and that can be devastating, no matter if it's your first-time starting seeds or your 100th. No one wants to put time and energy into something, only for it to fail.

Let me walk you through how to harden off your plants and avoid this shock. There's no need to feel intimidated; it is a pretty straightforward process. When it's safe to move your seedlings outside, you'll need to do this gradually and in increments. Otherwise, the harsh sun could scorch them, or perhaps you will have a downpour or windstorm that they haven't been acclimated to. Think of it as easing them into summer, just like you ease yourself. You probably wouldn't go outside on a 90°F (32°C) day after being cooped up in your house all winter without wearing sunscreen, right? Well, plants are very similar.

In the morning or evening, start to ease your seedlings into their new environment by leaving them outside for a few hours and then bringing them back into your greenhouse or house. The next day you'll continue this transition process, except you'll increase the amount of time that they stay outside and the amount of sun they are exposed to. This will continue until your plants have adjusted to being outside all day (and night) long in all weather conditions. Typically in 7–10 days, they will be ready to stay out in the garden.

I prefer to harden my plants off early in the morning, and set a timer on my phone for a particular amount of time so I don't forget about them.

Transplant time

Transplanting day is my favorite day. I've spent months planning, sowing seeds, potting up, and prepping, and now is the time to step out of my greenhouse and plant my seedlings in the ground.

There are only a few things you need to do to ensure success when transplanting out into your garden.

1. Follow the hardening-off process and your plants will be properly prepared.

2. Plant early in the morning or later in the evening to reduce transplant shock. Sometimes, even when adequately hardened off, plants will experience shock due to their roots being disturbed in the planting process.

If plants are healthy, they often bounce back quickly with no problem.

3. Water the soil both before and after you plant.

Watch them grow

Well, friend, you've done it! You have successfully started and grown seeds. This is when you give yourself a big ol' pat on the back and say, well done. I hope you found joy in starting your seeds and feel the excitement of getting to watch all your hard work turn into delicious and abundant food. In the next chapter, we'll talk about what happens next, how to have an efficient tiny but mighty farm, and the best tools for the job.

CULTIVATING FAMILY

**When we create space for our family to
join in on our passion, it ignites a power
they didn't realize that they had.**

Most gardeners remember when they started seeds for the very first time. There are a few memories that, as gardeners, we always keep with us. For me, this is one of those. I can tell you vividly about the day I started seeds for the first time. I was scared to death, but the excitement far outweighed my anxiousness. That day, I told myself I wanted to create this same experience for my kids. The following year, I extended the invitation for my children to help me start seeds in the greenhouse. Their responsibility was to seed, water, and care for those plants. They were invested in the success of their plants and stewarded them well. When it was time to plant the seedlings out in the garden, they were right by my side. I created a space for them. I invited them into my sanctuary. When we create space for our family to join in on our passion, it ignites a power they didn't realize that they had. Always extend an invitation for your family to be more involved with growing their food. Take the extra time to explain a variety's history or teach them how to prune a tomato. When you sow into the future, you are sowing the next generation of gardeners.

Chinese cabbage is a fun variety to grow. Along with carrots, peppers, and some garlic, this will make delicious kimchi in the winter.

CHAPTER SEVEN

Tools & Efficient Systems to Grow On

It's such a joy to share my journey with you through this book. I feel as though my heart has been poured onto every page, and through that, I've shared failures, victories, and hopefully good applicable advice to equip you to make your tiny farm a mighty force to be reckoned with.

However, this journey hasn't been easy for me, and it likely won't be for you either. You will often have to make hard decisions on your farm, whether they be about crops, growth, systems, setbacks—you name it. You will have to remind yourself daily of the importance of what you do. Hard choices will have to be made, but I'm strong enough to make those choices, and I know you are, too.

I tell you this because I want you to lean into this next part. I sure didn't at the beginning, and it cost me time and money, and now I want to help you avoid the mistakes I once made. Years from now, I want you to look back and see the pivotal moment that shifted your farm into one

of efficiency and effectiveness, not only for your family but as a teaching model for your entire community. I remember sitting down one day with my papaw and explaining everything I was doing on my farm, the systems I had put in place, the growth I'd experienced, and I could see the light sparkle in his eyes. I knew he was proud, and that made me even more grateful to have put in the sweat required to get my farm to where it is today. Not only was my farm efficient, but it was something I felt proud to tell others about. Now that we've discussed the why, let's tackle the how.

Establishing systems

Author Ben Hartman taught me a practice called "Muda" through his book, *The Lean Farm*. Muda is a Japanese word meaning futility, uselessness, and wastefulness and is a key concept in lean process thinking. Muda prioritizes waste reduction as an effective way to increase profitability. Ben uses the practice on his farm to cut out anything that does not add value. He emphasizes reducing waste and maximizing a lean approach. When reading his books, I was captivated by his efficient systems and the disciplines he uses on his farm.

When I finished reading Ben's book, my mindset shifted. I finally understood the weight of creating a system for my farm. I knew it would evolve over the years, but it could be the one thing that broke my farm if I did not make it a priority.

And so that was it, no more willy-nilly planting, no more do what I want, when I want. I was becoming drained by my setbacks, and it was all due to a lack of systems.

> **Muda is a Japanese word meaning futility, uselessness, and wastefulness and is a key concept in lean process thinking. Muda prioritizes waste reduction as an effective way to increase profitability.**

About Systems

Give yourself grace when establishing systems on your farm. It may not be second nature in the beginning, but before you know it, all your systems will be in place and will make your job a breeze.

I refer to systems as a set list of procedures we follow on our farm. It's organized, structured, and done the same way over and over again. For instance, we have a system for cultivation. Every Wednesday we cultivate our beds. It's never in question what we do on that morning because it's a practice we have put into place on our farm every week.

On Tuesday we harvest flowers. Every Tuesday at 6 am the crew is ready to go, and we start with the high tunnel, then work our way into the raised bed garden, and finish with our patch of sunflowers. We have this system to keep us efficient and organized. Instead of spending our morning wondering what needs to get done, and in what order we have this already mapped out for the day—which makes getting tasks done easier and faster.

CULTIVATING COMMUNITY

Strike up that conversation with a stranger, donate some extra produce, or maybe gift an arrangement of flowers. I promise you won't regret the relationships you'll make along the way.

When we think of community, often we think of the neighbor down the road or the friend we grew up with, but I'm going to prompt you to think differently. Community is wherever you choose to cultivate relationships. Most of my community isn't local. But they are a strong force that encourages me, motivates me, and propels me forward on the hard days. Seek out opportunities to connect with other farmers and growers through social media, comment on their posts, send messages, and be willing to create friendships. This is such a large part of my life and has been one of the sweetest ways I've met new friends and connected with like-minded people worldwide.

I am what one might call a free spirit. I love this about myself, but it proves challenging when sticking to a structure. Systems go against everything in my nature. After wrestling with hard conversations and my plans for the farm, I knew a shift was necessary for the growth I wanted to see.

What did I know about structure when farming? I thought back to my beginning days learning while working on other farms. Every excellent farmer I knew had a system they used to structure their farm. Looking back when I volunteered, one farm had certain days for weeding and planting.

Then, there were days when they harvested and made deliveries. You could always find their tools organized in the shed; even harvesting buckets were labeled and in their appropriate place.

Systems and structure are what makes us good farmers and gardeners—adapting, learning, and bettering ourselves and the things around us. Let's walk through the process so you can cultivate systems for your farm. In full transparency, I am still creating these systems, changing plans, and improving my farm daily, and you should be too. Does this mean I never willy-nilly plant? Of

My daily chore chart

Crop	Seeding date	Size	#	TP,DS,TC

Don't over complicate your chore charts. Keep them simple, task minded and easily readable for anyone else who may be watching your farm when you are gone.

A daily chore chart is a great place to start structuring the day. This chart is not a set of rules that can never be broken. Instead, this chart is a guideline to help keep me accountable for what needs to be done. Creating similar charts for your farm will give you a clear idea of what tasks you have to tackle on a daily basis.

Monday Most of my Monday mornings are spent in the greenhouse seeding (I do a bi-weekly succession of spinach and lettuce). These crops get seeded on this day and the prior week's starts get transplanted into the garden.

Tuesday I am often harvesting flowers and produce on Tuesdays for later deliveries.

Wednesday This day has been deemed *Weeding Wednesdays*, and it is also for cultivation and bed prep if needed.

Thursday Thursdays are my office days, for responding to emails, entering farm charges, paying bills, and handling farm expenses.

Friday On Fridays I walk the farm with my notebook and take into account what didn't get done and what needs to be added to the next week's list.

As mentioned, this chart changes based on the season, and living on a farm comes with its fair share of surprises—so there's adapting and adjusting with the seasons. This summer we are transitioning into dahlia production so this chart will shift with the demands and needs of that crop. I like to jot this information down on a large dry erase board I keep in my greenhouses. That way, the dry erase board chart is easy to access and change as needed. I find it helpful to keep visual reminders of what needs to be accomplished on your farm. It is helpful when keeping up with your daily to-do tasks but also for deciding what the vision and direction is for your farm.

course not, but I do far less of it, and its appeal has weakened greatly. I am learning to thrive with the structure that's been put in place, and I have already seen the betterment of my farm because of it.

Mentoring is an excellent way to see how other growers put systems in place, many of the systems I follow today are modeled after those on surrounding farms I've had to privilege to spend time on.

Create a vision

To have a system in place, you must first create a vision. When you visualize your farm, what do you see? Write it down, make a dream board, and talk about your ideas with other people. Get excited about the endless possibilities for your gardens and farm. When you build that excitement, your passion grows, and you will be more productive when putting these practices into place because you know what it is you are working toward.

I've always had this vision of being a pillar of hope and possibility, of growing food until my family's heart was content, and always offering encouragement to anyone willing to listen. I want to leave a mark that generations from now might come to know. I have no interest in having the biggest farm. In fact, I am more than content with my small-scale farm. But what I do have an interest in doing is making a big impact with my farm that will inspire multitudes. I want to embrace nature and integrate its work with mine. I have this dream of creating a domino effect and having a younger generation take hold of this lifestyle and run wild—encouraging even more people to take the leap and plant the seed, offering more organically grown food, and changing how we all think of gardening. Dream wild, beautiful friends, you deserve to leave a mark too.

> ### Dream wild, beautiful friends, you deserve to leave a mark too.

Farming is all about finding a balance. I never want to change who I am, so I find times when my free spirit can run wild and know that there are times when I need to get down to business.

Now that we've dreamed together, let's start planning together. Once you've established a clear idea of the model you want for your farm and the style in which you choose to grow (raised beds, in-ground, no-till, containers) you can start to dig in. You may be wondering why having a vision is such a crucial part of the planning process. Here's why. After you know your vision, you can start building efficient systems based on your goals and aspirations. This enables you to simplify your planning process and beyond.

Set realistic goals

Hi, I'm Jill, a wild dreamer, filled with wonder and the possibility of what could be. I believe there is a mighty power in dreaming and believing in what you can achieve, but I have found in gardening and farming that this journey of creating your dream farm is not an overnight process. *It takes years to get it right.*

Cultivating systems for your farm will increase
your yields and reduce your work time.

The appeal of planting without purpose fades greatly when you realize the lack of production it yields. If growing to replace the food needs of your family, production will ultimately have to take a front seat.

When we learn to set realistic goals for our farms, we stay encouraged. When we do not implement practices that move us toward our goals, that quickly changes. I fully understand how exhilarating it can be to sit down at your table with pen and paper in hand and dream of the wildest things, and while I will be your biggest cheerleader for dreaming big, I have found it best to set smaller, more attainable goals within the bigger vision. This is not only more practical, but it also helps you stay focused on the next step and not waste energy on steps that are years down the road.

I like to give myself weekly goals, monthly goals, and annual goals. This is helpful because it lays a clear foundation of what is a priority. It also accentuates the things I need to work toward to eventually achieve all my goals. As you continue using the systems I'll introduce in this chapter, they will become second nature to you. Your budget should also be considered when laying out your vision and goals for your farm.

First-year planning

If this is your first-year growing, don't tackle everything at once. That means finding only a few crops in the seed catalog to plant, starting with a couple of garden beds, and expanding every year thereafter. Learn to grow that handful of crops well, maximize your bed production, work within what you've created, and then expand.

Instead of feeling limited by what you are growing, lean into empowerment as you learn the ins and outs of your new systems.

Several years into our farming journey, we hired a farm consultant. He is a dear friend and my mentor; he helped me build a plan for my farm. These were some of the questions he encouraged me to ask myself in my first year of "real farming," and I will ask you to evaluate the answers to these questions for yourself as well.

Assess your goals (for your family and your farm)

1. How much time and energy do you have to grow your food and farm?

2. What do you want to produce on your farm?

3. How do you intend to fund these goals?

Assess your values

1. What are important values to you?

2. What practices do you intend to put in place that align with your values?

3. Are there things that are high in value for you but do not directly benefit you? What are simple ways to accomplish them?

Knowing your skills and assets

1. Do you have any experience producing at the scale you intend to reach?

2. Do you have a mentor or a community you can go to for advice and guidance?

3. What is your skillset? Can you build and manage the things you are trying to achieve? If not, do you have resources to help you achieve them?

When you start your week evaluating everything that needs to be done on your farm, you have a clear expectation of your tasks throughout the week.

Five-year planning and beyond

After we asked ourselves these questions, we built out our five-year plan. That plan involved things we needed to learn, what we needed to build and buy, and the time frame for these things to get done. It was overwhelming at times, but we had a plan and a system, and it felt good.

We were taking the next best steps for our farm. Yes, we had our long-term goal, but we focused on the next small achievable goal that would get us there.

I say these things to you and share our journey openly because I genuinely want you to feel encouraged by this process. Your journey will be intricate, unique, and built to fit your family's needs. Embrace the season of planning, and you will be in the hustle and bustle of reaping the fruits of your labor for years to come.

What we teach the next generation will change how they view food. My prayer is that we don't take that lightly. That we instill the value of its importance and make a mark that will forever impact them.

Reducing wasted time & resources

Established pillars within your systems are the things you continuously fall back on and ask yourself when implementing something new on your farm or to help you nail down efficiency with current practices. These are our three main pillars.

1. Establish steps that add value (in terms of your time or goals)

2. Use tools that add value (make your job easier)

3. Cut out anything that does not add value to your farm (practicing Muda)

Look back on these pillars before making decisions. It helps you see with a clearer mind.

How do we reduce waste on our farms? Get an outside set of eyes to look at the issue. I have found this an extremely helpful way to cut back, especially on wasted time. Ask yourself, am I spending two hours planting when I could put a system in place using tools that add value and get the job done in half the time? If the answer is yes, then clearly, I need to do some shifting.

We all want to reduce wasted time, but often we don't realize it's happening; this is when it is helpful to have an outside opinion. Be willing to listen to what they have to say and make changes as needed. A great way to go about this is to have your farm consultant spend the day with you on your farm as you work at normal tasks, while they are noting ways to simplify the process. This can be as simple as tips to make seeding in the greenhouse faster or a more efficient way to hold rubber bands to wrap flower bouquets faster. These changes may not seem like huge adjustments, but often it's the small changes that make the biggest impact.

If your goal is to start a small kitchen garden, plant a few herbs that are easy to manage. If you want to grow food for your family, plant food you eat often. When you don't take on too much, the garden becomes more manageable.

Know which crops will need to be started either indoors or in a greenhouse and which ones you are direct seeding. Growth will vary greatly when you start or sow them outside.

Greenhouses and infrastructure are a higher cost item on our farm and require more thought to budget accordingly. If you know ahead of time what big ticket items your farm needs, you can start to create a plan.

What is a farm consultant?

Farm consultants are also known as agricultural specialists, and they help farmers and growers of various size scales to evaluate their growing space and learn how to maximize efficiency based on their goals. They can also review financials and aid in potential business growth.

When learning new systems and ways to reduce waste, it is also helpful to get rid of things like tools that take too long to repair or aren't making your job easier. This can even apply to preserving your harvest.

I used to only preserve my harvest by canning, which is fine, but now it is not my primary way to preserve the harvest because it takes up much of my time. When I thought about the time I could spend outside growing more food instead of standing over the stove all day with pressure canners, I changed methods. Now, I freeze-dry and vacuum seal most of our harvest, allowing me more time to be out in the garden, producing more food.

When I was challenged to ask myself the same questions I asked you above, I had to really take

my time and consider the practices I had put in place. And even to consider the practices I had not adopted. Was there room for growth? You bet! After our eight-hour day was coming to an end with our consultant, I felt heavy yet hopeful. We had a clear idea of the changes we needed to

If you start the majority of your seeds in a greenhouse, finding the fastest way to do this will save you loads of time in the long run. Have all your trays filled with soil and ready to go first, that way when you get ready to seed, you can do this quickly because you have already filled up your trays.

make, the amount of income we needed to bring in, and the systems that had to be put in place to see these things come to fruition. If you meet with a consultant, and they have a lot of feedback for you and your farm, don't be overwhelmed. Start a list and prioritize the things you can accomplish first. New systems are not an overnight transition. Some may take months to put in place, while others may take you a year or more. Work within your means and keep moving forward every day. I truly believe that that is where we find success, by not overloading ourselves with long lists that never get checked off, but rather by focusing on a few things that are manageable and within our reach. Remember, you got this!

Finding the right tools for you

I would have never imagined I would be sitting in my greenhouse writing a chapter in my book about the difference a good tool can make. I feel as though my papaw would be proud. So, papaw, this section is for you.

Having the proper tools on your farm will, without a doubt, speed up tasks, and can all around make your job easier. When choosing the right tools for your tiny but mighty farm, here are some considerations to keep in mind.

Quality

You can spend so much unnecessary money on the newest tool to hit the market, but it may not even be necessary. It may also be poor quality. In fact, many of the tools you will find at box stores aren't high quality. They usually break after one season, and they weren't created with the grower in mind. Before buying a tool, ask yourself what the qualities of a good tool are. These are the qualities I look for.

- Functionality
- Easy to utilize
- Appropriate for your scale
- Durability
- Cost

You might be surprised how few tools you need to farm efficiently. It is not about having the latest and greatest, but it is about having versatile, long-lasting tools.

Don't get me wrong. I will always turn and grab my African woven harvest basket when taking an evening stroll through the garden, but on the days that we are just harvesting, I need good snips and a big, sturdy harvest bucket that's easy to handle to get the job done.

Functional tools are so important. It is essential to remember your values. For me, that means a focus on efficiency, but I also leave room for the things that make my heart smile, like my beautiful African harvest baskets or wooden handled trowels. You can find a balance between efficiency and beauty and functionality when you have sound, efficient systems in place. I am all about finding that balance; I enjoy the aesthetics of tools, baskets, and the layout of my garden in addition to wanting them to be functional and efficient.

> **Tiny tip**
> You will free up time when you invest in the proper tools for your farm.

I enjoy the aesthetics of tools, baskets, and the layout of my garden in addition to wanting them to be functional and efficient.

Because of this, I always leave a little room to go rogue. Most of my tools are versatile and make my day-to-day tasks much more manageable.

My most-used tools

A Modesto heavy-duty knife is one of the most used and most versatile tools on my farm. It is used for everything from harvesting cabbages, cauliflower, and broccoli to knocking down sunflowers on harvest days. It is durable, easy to use, and has lasted years.

Hori-hori knife
These hori-hori knives are extremely versatile. One side is serrated and can be used for cutting through root systems, the end is pointed and is used as a garden trowel, and the measurements engraved along its length, can be used to gauge depth and spacing when planting. This is a 3-in-1 multipurpose tool, and I use it daily.

Pruners for flowers and vegetables
Quality garden pruners are a must-have for so many tasks from harvesting flowers to pruning tomatoes. Investing in a good pruner will make hundreds of day-to-day tasks easier.

Harvesting knives
I have several harvesting knives for various things. I use a small serrated hand-held knife to harvest spinach and lettuce. Then, we have larger knives that are used to harvest heads of cabbage and broccoli. Investing in a few knife sizes is handy during harvesting season. To keep our knives from getting lost, we added a magnetic strip to each greenhouse and high tunnel so we always know where they are. This makes harvesting days a breeze.

Gardening tool belt
I didn't realize how much I would utilize a gardening tool belt. It is made from leather and has multiple pockets. One pouch holds my permanent marker and a small knife. The second pouch is bigger and holds a set of pruners. The largest pouch fits my phone. And the side has a ring that holds rubber bands for making flower bouquets. Before I had my gardening belt, I was constantly trying to fit various tools into my pockets and would always leave things behind. Having it all in one place, right on my hip, makes accessibility a breeze.

Broadfork
Use a broadfork for bed prep to simplify and speed up the cultivating process.

Cultivating hoe
The hoe I prefer for cultivation is called a stirrup hoe. It is so simple to use even my kids get to help out on weeding days.

Harvesting crates and buckets
The style of the harvest containers you choose will depend on what you are growing, but I prefer stackable crates. They work great when harvesting in a tunnel or out in the garden.

Tools for scale

If you desire to scale your tiny farm into a market garden here are some additional things to consider.

- Soil preparation
- Seed starting, preparation and planting
- Cultivation
- Harvesting
- Wash and pack systems

For those who intend to scale their farm, the things listed above will need to be thought through carefully. Additional costs will be involved with many of these items, especially a wash and pack system. The tools you need for growing using a market-style approach are different and may be more costly. Many of the tools I use are intended for the small-scale grower. As you begin to scale, your tool needs will shift as well.

We've covered why systems are critical and how appropriate tools can make or break efficiency, now let's transition into what those systems look like.

Getting organized

I have found through my years of growing that the backside of efficiency is organization. Being organized is a key to being efficient.

Freeze drying or dehydrating peppers is great way to make your own chili powders. In the winter when soups are a must, I can take out these seasonings I've created from the summer garden and continue to use them.

You don't need a lot of tools to be effective on your farm. In fact, a few small-handled tools and a couple of long-handled tools are more than enough to get the job done.

Examples of organization in action

Let's start at the beginning and take a look at some simple ways to be more organized.

Seed storage

If we get organized first with our seeds, this sets us up to have a successful seed-starting season. One way I do this is by buying large shoebox-style photo storage containers from craft stores. Mine has a handle and clamp to keep it closed. Within the box are individual clear containers (originally designed to hold 4 x 6 photos) that lift out. I store my seeds in these. Due to the excessive amount of seeds I have, I need to be very detailed with how I label each box. For tomatoes, this is broken into types: greenhouse tomatoes, yellow tomatoes, red slicers, etc. My seed containers stay well-organized at all times which saves me time when looking for varieties.

To take organization a step further, I break seeds down even more just before planting season. Instead of grouping my seeds by vegetable (cucumbers, broccoli, etc.), I group them by planting date. Usually, I have one day when the house ends up being chaotic as I organize this. Seed packets

Over-the-shoulder buckets are some of my favorites. When harvesting large amounts of tomatoes, peppers, and eggplants, these can get heavy fast. Having good back support to hold these buckets makes a world of a difference.

Multi-purpose plants

Implementing systems for seed starting is one way to max out efficiency easily by staying organized. When I am planning my crop rotation and what I will be growing for the season, I also think about what will serve more than one purpose. I grow varieties that help with weed management, and I also grow varieties that put nutrients back into the soil. Almost everything I plant serves more than one purpose. In our sunflower patches, I line the perimeter with squash and zucchini. For our dahlia production in the summer, I plant single rows, and on either side of them, I plant bush beans to act as a trap crop. The goal is to lure pests to the beans and keep them away from the dahlia petals. Not only does it serve as a pest prevention, there will also still be plenty of beans for my family to enjoy as well.

WHAT IS A TRAP CROP?

A trap crop is a crop planted to attract pests and lure them away from the nearby crop you intend to harvest. By planting trap crops you are reducing the risk of insects interfering with your main crop. It also reduces your need to use pesticides.

Watering

A few other efficiency components I put in place were increasing the number of frost-free hydrants we had on our farm. I added two additional ones to the raised bed garden, one in my seed-starting tunnel, and another in the high tunnel. By doing that, I eliminated the time needed to walk water hoses from one yard hydrant to another. This may seem like a small act, but when I decided to emphasize reducing my wasted time, I knew this needed to be done. As mentioned in previous chapters, I also placed drip irrigation in all my growing areas and have the system set on automatic timers for watering. This saves me more time than I ever thought it could.

I remember in my first garden, and even in my second and third gardens, I would spend hours

are spread out all over the living room floor, and I have my large dry-erase boards close at hand.

I start by grouping together what needs to be started from the beginning of February all the way into April. This is based on an "every Friday" sowing system. Every Friday, I reference the dry erase board for how many seeds I need to start, and in which type of containers I will start them. The board also states which varieties are to be direct seeded and which are to be transplanted, and where they will be planted on the farm; raised bed garden, high tunnel, cottage garden, etc. I also color coordinate the board; flowers I write in pink marker, vegetables in brown, and the things I need to grow to sell are in blue. This makes seed starting days much more efficient.

outside hand watering the entire thing, and while there was some tranquility to that, it certainly did not make sense as I began to scale and grow my farm.

Tool storage

Earlier I mentioned that the farm where I used to volunteer had a place for everything. That farm taught me so many practical ways to implement better organization. Aside from having an area in each tunnel with the tools I use most there, I designated my "greenhouse on the hill" as my tool shed. This was a cheaper greenhouse, made of materials that would not withstand the conditions of my climate. But it had potential for a new use. Lined with shelves, a decent roof, and plenty of space, I keep all of my long-handle tools in there now. I also have several totes, each labeled with various farm equipment. One is for irrigation supplies and another has row cover and insect netting in it. One tote has flower nettings and trellising supplies. When I have a need, I am not rummaging through the garage or other places on my farm to find them. It is all in one spot, organized in a designated tote that I can bring with me to wherever I am working, and then easily take back to the tool shed when I'm done.

Vertical growing

Recently, I started evaluating the amount of time I was spending on my trellising systems. In the past, I would use cattle panels mounted on T-posts to trellis beans, tomatoes, and other various climbing plants. My garden was lined with these arch trellises, and you can still walk out in my raised bed garden now and find squash dangling from them and nasturtiums climbing up them.

But, for the crops I needed to maximize production on, I knew there had to be a better system. In our high tunnel, I phased out the use of cattle panels for a different trellising option and this is why.

1. I could not put up the cattle panel trellising system by myself. I would have to wait for my

If you try a few buckets and don't like them, try another. It took me a few seasons to finally land on one that was functional and met my needs.

husband to come home in the evening or put it on the weekend to-do list. This was costing me time, and we all know the saying "time is money."

2. It was costly and labor-intensive. As we are seeing a rise in materials cost, cattle panels and T-posts are continuously getting more expensive each year. What would have cost me $40 for a quick trellis is now double that, and as I continue to expand my production, this was a cost that could easily be reduced by implementing other types of support systems.

Now, I use jute twine that I buy at my local feed store for a few dollars. I fasten it to the top of my high tunnel, using a 2-leader system. This is the way I now trellis my cucumbers and peppers. It only takes me a few hours to have everything done and in place.

Day-to-day rhythms

The last thing I want to touch on when it comes to staying organized is to find ways to integrate your farm with the natural rhythms of your day-to-day tasks. If you are considering selling produce or cut flowers, think about how you can save time and money on deliveries.

If that looks like taking your children to sporting events or maybe it's your weekly day to head to town for groceries, try and coordinate all of your deliveries for a time when you are already making a trip away from the farm. I try to keep the days I spend off the farm limited, and how I do that is by having one or two days when I am going to town to get groceries, taking my girls to gymnastics, or heading to the gym. I choose those days and coordinate them with my deliveries to efficiently maximize my time.

USING A 2-LEADER GROWING SYSTEM

With a 2-leader system you prune and train your plants so that they have two main leaders growing upward instead of one. I do this when my plants are still in the greenhouse and have put on their first set of true leaves. At this point, simply take your pruners and cut off the top. By doing this your plants will focus their energy on the two leaders that are left and will continue to grow this way. There is a difference between a leader and a stem. A leader can produce fruit, a stem cannot. Once we transplant these out in the high tunnel, I take twine from the top of the tunnel and attach it to the bottom of the plant so that both leaders will each continuously up a twine trellis as they grow.

I keep dry-erase boards in every tunnel. This one is in the glass greenhouse, I have another one hung in the seed-starting greenhouse that displays succession sowing I need to do, and one smaller one in the high tunnel to calculate yields.

Time management

Good time management may seem easy, but that's not always the case. It is easy to let time get away from you when you are caught up in the wonder of the garden. Blocking my time out on a calendar is the most helpful way to stay on track.

By establishing healthy time management skills, you allow yourself more freedom to invest your time back into your farm or into hobbies you enjoy.

Creating a schedule is one way to help with this. For instance, on Wednesdays, I cultivate (remember Weeding Wednesdays?). I know every Wednesday from 9-11, my time is blocked off for that particular purpose. On Fridays, I evaluate the next week's

to-do list, on Tuesday I harvest, and so on and so forth. Each day has specific time blocked for a designated purpose. When creating a schedule, set dates and times for tasks you know you need to accomplish.

When you understand the importance of setting a schedule and defining tasks for yourself, you can note your accomplishments by marking things off your to-do list. It feels good to know you are moving forward.

Once you've maxed out your time effectively and put systems in place that make your job easier, you have will have time to explore other things and will be able to spend time giving back.

Giving back

One of the best feelings about establishing rhythms is the opportunity to give back. Giving back is pouring your knowledge and experience into someone else. It is very fulfilling to mentor someone so they can learn to ask the hard questions, put in the necessary work, and reap the rewards.

I genuinely believe we are better together than we are apart. Leaving our land better than we found it, implementing sustainable practices, and growing organically in a way that isn't harmful but works in harmony with nature are ways to work together with our community. As small farmers, we are building a community of like-minded people who sustain their families and learn to steward this Earth well. Together we are all moving forward for the greater good.

> **I genuinely believe we are better together than we are apart.**

Every person you meet provides an opportunity to share your values. Explain to them why you do what you do. Share that conviction you feel about the need for growing healthy food—the heaviness we carry because of our broken food systems and the urgency to do something about it. Explain to them some of the systems you've put in place on your farm to help speed up essential tasks or perhaps make harvesting days easier. I used to always think "Well, people probably already know that so what's the use in sharing?" But now I know that is not true, and I've built my life and my farm around the ability to teach others.

Invest in your knowledge

I have never been more eager to expand my knowledge than now. I want to read every book, listen to every podcast, and go to farming and homesteading conferences. Anything that will help me be better and do better.

You've heard the saying "knowledge is power?" Well, it isn't wrong. The more you know, the better you'll be at growing on your tiny but mighty farm. The more we understand why and how to grow food, its impact on our families, our local economy, and most importantly, our planet, the better versions of ourselves we will be.

I cannot encourage you enough to pick up that farming or gardening book and read it, take that online course, or go to that workshop. You will not regret pouring back into yourself and gaining knowledge to make your farm the best farm it can be.

Grow on, gardeners, and buckle up for the next step on our ride—a look at how growing structures like high tunnels and greenhouses can make your tiny but mighty farm even stronger.

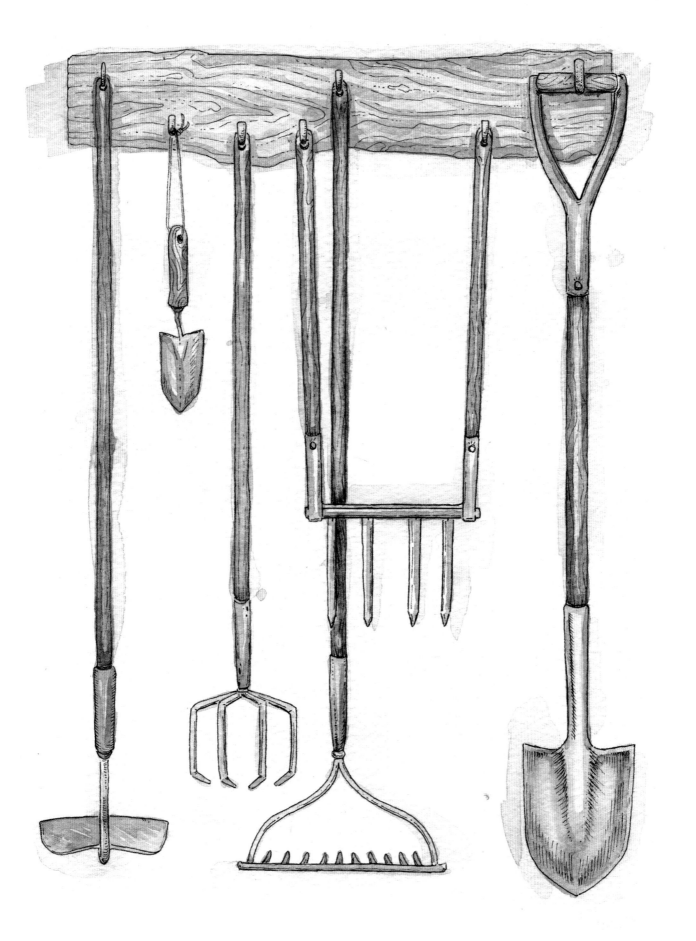

For varieties like pole beans, vertical trellising i
a great support, makes harvesting a breeze, and can
be used season after season.

CHAPTER EIGHT

How Structures, High Tunnels & Greenhouses Help

A few years ago, we had an unexpected winter storm in our forecast, with single-digit Fahrenheit temperatures predicted during the day and talk of an intense ice storm coming right for us. We ended up having a record-breaking snowfall that our state hadn't seen in over 100 years. We received around 14 inches (36 cm) of snow in a matter of 48 hours. For us folks in the southern U.S., that's a lot.

I remember it like yesterday; we were at a pig butchering workshop at our friends' farm for the weekend (the farm we now own). A few other couples were there, along with the two gentlemen who were hosting the workshop. What better time to harvest pigs than in the middle of a crazy snowstorm, right?

I'll spare you some of the details and cut to the good part, but the snow kept falling and the roads were not safe to drive. So, we were stuck; all of us. The first thing that came to mind was the relief that a lack of food wouldn't be an issue. Aside from the two massive pigs that had just been harvested, my friend, a master

gardener, had a high tunnel packed with food.

I bundled up in my 14 layers like any amateur in my situation would do and shuffled my way down to her polytunnel. I opened the door and was immediately greeted with warmth flowing upon my cold, red cheeks. I was shocked at how warm it was. Sure, I had been in her greenhouse in the spring and summer and saw the massive impact her high tunnel had on the vegetable production, but I hadn't seen it in action during such a harsh winter.

Nothing was phased; the harsh weather didn't affect any of the plants. The kale was still vibrant and green, the cabbages were lush, and the rutabagas were protruding from the soil's surface. It was like summer (well, almost) in the dead of winter. The plants were thriving as though they hadn't received any of the snow that fell on the raised bed garden. If they weren't in the greenhouse the weather would have stunted or killed the plants.

I harvested a basket full of vegetables and herbs and headed back to the house. I knew right then and there how valuable season extension was to her farm and soon would be for mine, too.

Extending your season

Later in this chapter, I'll discuss the different types of season-extending structures and their roles on a farm, but first, let's look at what season extension is and how it is achieved.

Generally speaking, season extension is any technique or design that allows crops to be planted earlier or live longer than they would if growing unaided and left unprotected. A greenhouse or a polytunnel/high tunnel is usually what comes to

Types of season extenders

If you are having difficulty remembering what each structure is, and how it traps heat, here is a simple way to look at it.

- A greenhouse is a structure covered in glass, plastic, or acrylic that acts as a heat sink. It will warm up faster and stay warm longer.

- A row cover acts like a blanket.

- A polytunnel/high tunnel is a tunnel typically made from a steel tube frame and covered in polyethylene. It is usually semi-circular, square, or elongated in shape. The interior heats up because incoming solar radiation from the sun warms plants, soil, and other things inside the structure faster than heat can escape.

With a trellis system like this, you take a tomato clip and attach it to part of the plant and part of the twine. As the plant grows you continue the clipping process until the plant has reached the top.

people's minds. The sun's energy passes through the plastic or glass structure and is trapped inside, heating the structure. Row cover, a lightweight, spunbond, polyester fabric, is used much the same way but on a smaller scale. It can also be deployed in the evening to help trap heat around the plants and soil. It acts like a giant blanket during the night. These techniques allow plants to be better protected from the elements, especially during sudden cold snaps.

The lovely thing about extending your growing season is that anyone can do it. Sure, some folks may need to think about season extension earlier than others, and some may live in regions where cooler weather isn't a threat, but for most of us, we will have a winter of some sort. With season extension in place, winter is just another opportunity to grow food.

Protection from the elements

I have such a love for high tunnels. They have played a huge role in how I grow food and flowers on my farm. They have extended my season both early in the summer and well into winter. But that's not the only reason I've been thankful to have them lined up in my backyard. The elements are unpredictable for many of us. For some, you may experience harsh snow and ice. For others, threats

I truly believe if we all make one small positive change, then together, we make a much larger impact. By growing your own food and supporting local farmers you are keeping your dollars local and investing in your local economy.

Create a space your family wants to be in. If they only associate the garden with work and chores, they won't enjoy spending time there.

of excess rain may be an issue. For me, I have all those conditions and can also add intense hail to the list of weather reports coming from Arkansas.

A few weeks ago, we had threats of tornados in the area. I had thousands of plant babies lined up in my seed-starting tunnel. We participate in a large plant sale every year with other farmers in our area, and most of these plants were intended for this sale, and of course the week before the event, the storms started to blow through. Curled up in my basement with my husband and daughters, it sounded like a war zone was taking place outside. I asked my husband, what was making that sound? His reply was "Hail." My heart sank. I had just transplanted the first round of flowers into the raised beds and there were thousands of thriving plants started in the greenhouse. My mind started to wonder, and naturally I assumed the worst. By the time the storms had passed, it was too dark to assess any damage. Bright and early the next morning, assessing the damage was the first thing on my list.

As I walked into my raised bed garden, my heart felt heavy. The daffodils at the entrance were destroyed, and the transplants that just went out looked as though someone took a string trimmer to them. My brassicas were demolished. It was such a heavy feeling knowing I would have to redo all the work that had already been done.

High tunnels can be used in all four
seasons and make growing food year-round
more attainable.

Our 14 x 36 feet (4 x 11 m) greenhouse is lined with tables on either side and one down the middle. The front of the greenhouse has a large trough for soil and an area for racks that can hold tools and trays.

As I headed down to the high tunnel, I didn't have much hope. The closer I got, the more holes I began to see. Just imagine what a plastic-covered tunnel would look like if someone took a shotgun to it. Well, there you have my seed-starting tunnel, newly built, but boy had it seen better days.

Now before we all get ahead of ourselves and assume the worst like I did, there is a good part to this story. While the plastic on the tunnels looked, well, rough, the plants inside were lush, green, and full of life. Nothing was damaged. The plants that had been started for the sale were thriving. The flowers in our large tunnel were still blooming. Nothing was phased. Previously when someone asked me how I liked growing in my tunnels, my response was always "I love them. I can grow anytime of the year," but now after the hail and windstorm, when someone asks me that question, I will have new praises to sing about my high tunnels.

Had all those plants been outside under row covers, they would not have made it. Had I transplanted them out into my raised bed garden, they would have died. The only reason they were thriving is because they had protection from the elements, and goodness, am I thankful for that.

When reading through the different styles of structures available in the next section, consider this as well: it's not only about extending your growing season, it's also about protecting your plants from life's unpredictabilities.

Growing with the seasons

Have you thought of a greenhouse or a high tunnel as a super-insulated structure with a heat and power source? What if I told you supplemental heat and a power source aren't necessary to grow plants through the winter? Would you believe me?

Well, folks, I am here to tell you it can be done. We achieve this when we quit trying to grow heat-loving plants in the dead of winter and grow cold-loving plants instead. There is a wide variety of cold-loving crops that thrive in cooler/freezing temperatures. These are the varieties we should be growing in high tunnels and under row covers through the fall and winter. The same can be said for summer high tunnel production. Certain varieties thrive in the high humidity of a high tunnel and are great for greenhouse and high tunnel production.

When we grow with the seasons, what will thrive during each different time of year should be at the forefront of our minds. The last few years we have ventured into cut-flower production as a new stream of revenue for our farm. I will be honest, I love it. Sometimes even more than growing vegetables. I tell you this because many flower farmers grow in high tunnels. When you are evaluating your farm goals and what you want to achieve, the picture in your mind might look different than having rows of tomatoes. You may envision fields of sunflowers, and high tunnels full of snapdragons.

What to grow in high tunnels for winter production

My favorite cold-loving crops to grow in high tunnels

- Curly kale
- Spinach
- Brussels sprouts
- Cabbage (Chinese, red and 'Tiara')
- Carrots
- Beets
- Rutabaga
- Broccoli
- Kohlrabi

Some flowers also do exceptionally well when grown in low tunnels (also called caterpillar tunnels), high tunnels, and even overwintered with the use of row covers.

My favorite flowers to grow in high tunnels over the winter

- Carnations
- Snapdragons
- Dara
- Pincushion
- Sweet peas
- Nigella
- Orlaya
- Bachelor's buttons

Setbacks are going to happen on your farm; how you handle those setbacks will determine the outcome. Don't stress over things you can't control. Replant, and keep putting one foot in front of the other.

The options for various winter vegetables and overwintered flowers are endless. Be mindful of your growing zone and the conditions you are choosing to plant in before deciding on which seeds are best for you.

PLANT DORMANCY

Some of the crops you might choose to grow throughout the winter may not continue to grow in January and February (for me, this is cabbage and brussels sprouts). The plant just stops producing and goes semi-dormant during those super cold months. Thankfully, you are still able to harvest from the plants even though they are not actively growing. Although my brassicas quit producing new growth, I was harvesting them through the winter and well into March. Think of your high tunnel as a giant walk-in cooler for your plants. It's excellent storage, and you can harvest fresh as you need.

When we expanded into growing during all four seasons, you might be surprised to hear that we reduced our workload tremendously. This is because extending your harvest also removes the need to put up and preserve as much food because you will continuously harvest through all four seasons.

I always found it interesting that growing a garden was only something we did in the summer. Growing food in the winter was like a foreign world to me. Now I look back and wonder why I felt that way. If we could grow food in the summer, why couldn't we grow food in the winter?

These days, I often get asked, "Isn't it more work to garden year-round?" to which I reply, "Isn't it more work not to?"

> These days, I often get asked, 'Isn't it more work to garden year-round?' to which I reply, 'Isn't it more work not to?

The thought of eating out of season, making unnecessary trips to the grocery store, and a garden with no food is depressing. In all honesty, it takes no more time out of me to garden year-round using my high tunnel than the extra time I'd have to spend preserving all the harvests of summer.

As you can see, there are many reasons to contemplate extending your season. I like to think of the seasons as complementing each other, not competing. There are particular varieties I can only grow in certain seasons or I risk not having access to them at all. Root crops take on a sweeter note when grown through fall and winter; lettuce doesn't become bitter, and brassicas thrive.

Despite the hail damage seen in this photo it did not affect my plants. I will repair the bigger holes with plastic and tape, but the entire sheet of plastic does not need to be replaced.

Our large high tunnel does not have heat or fans. In the winter we keep the side walls rolled down to trap the warm air in, and in the summer we roll the sides up and place a shade cloth on top to keep the temperature down.

Root crops take on a sweeter note when grown through fall and winter; lettuce doesn't become bitter, and brassicas thrive.

Aside from the freshness and excellent taste of growing your food year-round, it is much more nutritious to eat from your backyard than from food shipped in from many miles away. Did you know that vegetables begin to lose their nutritional value over time? Spinach, for instance, loses 90% of nutrients within the first 24 hours after harvest.

You cannot beat vegetables fresh from the soil to your plate versus the salad you'll find sitting on a grocery store shelf. They may look the same, but we all know the taste and quality are considerably different.

Another benefit of utilizing season extension is the variety that's available to you. You can grow your food with the seasons and allow your cooking to reflect that. Sautéed cabbage and root veggies in the cast iron skillet are a winter luxury. Fresh sliced tomatoes and cucumbers are refreshing after a hot summer day. Freshly picked salad greens and roasted butternut squash become a fall staple.

Each variety reminds us of a different season and tastes better when grown within its preferred season.

Do you want that freshness for your family even in the *off-season*? If you answered yes, then the rest of this chapter is for you. I will walk you through different styles of greenhouses, high tunnels, and their purposes.

Season-extending structures: establishing the differences

Before we dive deeper into the pros and cons of each different season-extension structure, let's first establish a few differences.

Traditionally greenhouses tend to be used for more commercial operations, and they are often anchored to the ground.

A high tunnel uses **passive** ventilation for air exchange and cooling, and a greenhouse uses **active** ventilation.

A greenhouse, by definition, is a "glass building in which plants are grown that need protection from cold weather." In most cases, greenhouses are equipped with electricity and have automated heating and ventilation systems. A straightforward way to remember this difference is *high tunnel=passive* and *greenhouse=active*.

Cold frames are quite different from both greenhouses and high tunnels. These have no permanent foundation and are typically smaller and closer to the ground. In most smaller cold frames, there are no ventilation systems in place.

We will explore many different structure options in a bit, so if you do not already have a structure for season extension on your farm, pick up your notepad and start taking notes about which one of these structures would suit your needs best.

CULTIVATING SELF

If structures are not in your foreseeable future due to budget, goals, or resources, don't get discouraged. I grew food for years without the use of any of these structures. I started seeds in my house without the use of a greenhouse, and only grew food during the summer. If that is you, that is okay. One way I kept myself from feeling discouraged was by focusing on what I was doing well. Celebrate the victories you are achieving on your farm and in your gardens, no matter how small. Doing so will also celebrate you!

The flowers we grow in our high tunnel are overwintered so they bloom first thing in the spring. We typically start harvesting around March and will continue well into June and July.

Caterpillar tunnel

Caterpillar tunnels are an affordable way to achieve temporary covered growing space. These simple, tubular steel structures are fixed in place with simple ground anchors. There is no digging or drilling involved. You can take the tunnel down, move it, and reuse it from season to season. Among the many season extension structures used on vegetable farms today, the caterpillar tunnel is one of the cheaper options.

Often these are also referred to as three-season tunnels. Most growers utilize them in the spring, summer, and fall, and then take them down for the winter. They are a structure that isn't built to withstand harsh winters. For those growing in southern climates, you can utilize these for all four seasons, as the likelihood for a large amount of snow and ice isn't an issue.

Benefits

Caterpillar tunnels can be built in various sizes, which is nice because essentially, you can make them however small or large you'd like. One of the appealing benefits of this structure is the

Caterpillar tunnels are an inexpensive way to maximize yields on your farm, especially if you are renting and can not put up permanent structures.

low-cost investment. Unlike traditional hoop houses, caterpillar tunnels secure the plastic using nylon rope that is zig-zagged over the top of the tunnel between each pole, creating a caterpillar-like appearance—giving it its name. Unlike hoop houses, you aren't using ground boards, hip braces, or other costly materials.

Due to its somewhat ease to take down and move, you can rotate caterpillar tunnels around crops. You can disassemble them in the winter if you have threats of bad weather. Or if you are renting land, they offer a viable option to expand your growing potential without investing in a permanent structure.

Whether farming flowers or vegetables, you can get a head start on the growing season by warming your soil and planting earlier when using caterpillar tunnels. If you desire to build soil around your farm, having this mobile option is undoubtedly a plus. However, if you choose to use a caterpillar tunnel in the same spot from year to year, you can also secure it more permanently by using materials like rebar and ground posts to hold it in place.

These tunnels are an excellent option for beginners who want to become familiar with growing through more than one season. While this may not be something you are used to, having a cost-friendly opportunity like this allows you to expand and increase your knowledge while working through the ins and outs of growing in a covered structure.

And lastly, caterpillar tunnels offer a kickstart to the season. We tried our hands at growing tulips this year with some friends, and we were able to harvest tulips a full month ahead of when other growers had them available for the market. We achieved this by coordinating the timing of the cut-flower harvests with other growers so that everyone's methods were timed for a specific target of the local consumer market. We were growing the bulbs in crates housed in a caterpillar tunnel and ended up making the first harvest by a long shot. Another farmer grew some tulips in-ground under row cover and had the second round of cut tulips available. And then finally, the farmer growing in-ground without any cover had the last round of tulips available. It was a win-win situation for the consumer and all of the farmers.

In the next chapter, I will dive into what it looks like to grow with your community and the aspect of selling to market. Our tulip story, just mentioned, is an excellent example of how you can partner with the growers around you, meeting everyone's needs to serve your community in the best way. It's crucial to remember that it's not a competition between your farm and the growers around you.

It's more about learning what your farm specializes in and learning how to intertwine that with other growers in your market area.

Considerations for the long term

There are a few concerns with using caterpillar tunnels, the most important of which is the weather. If I say it once, I'll say it again and again, the weather is always unpredictable and possibly one of the biggest challenges for any gardener. While it is intriguing knowing there is an affordable option to extend your growing season, caterpillar tunnels are more prone to wind and snow damage than other options—like high tunnels—because they lack stability since they are not anchored to the ground in the way that high tunnels and other greenhouse structures are.

> **If I say it once, I'll say it again and again, the weather is always unpredictable and possibly one of the biggest challenges for any gardener.**

If your lettuce tastes bitter, try watering more frequently or growing it through the fall and winter. In my zone, our springs are too hot for lettuce to thrive, therefore I am only planting these during the fall.

Even the plastic is secured with a rope instead of metal channel wire, a gust of wind or a crazy storm can send your plastic tunnel covering blowing down the road—leaving your plants damaged and sometimes wholly ruined.

You can avoid this by finding a location that offers some protection from harsh winds or by making your caterpillar tunnel a more permanent structure by adding purlins or wooden bases. The main downside of this is that if you choose to make these improvements, you eliminate your tunnel's mobility and increase your cost which are two of the most enticing features of these tunnels.

Low tunnels

Low tunnels are temporary structures made from half-hoops of various materials such as heavy gauge wire, metal conduit, or PVC pipe, all covered in plastic. Low tunnels are typically between 2–5 feet (61–152 cm) tall. Much like the row cover I've mentioned before (and will discuss in more depth later in this section), low tunnels are a cost-friendly option that aid in season extension and pest control. The idea isn't necessarily to extend summer production but rather to limit exposure to freezing temperatures for your plants throughout the winter and into the early spring. Low tunnels are effective because they trap the heat inside and keep out wind, rain, and snow.

High tunnels

High tunnels are becoming increasingly well-known due to their high return on investment and ability to extend your growing season either through early- or late-season production. But what is a high tunnel? Is it a greenhouse? A hoop house? I still get confused sometimes about the correct jargon around all these different structures!

The term high tunnel often gets used rather loosely, but high tunnels were originally greenhouses with high sidewalls that would allow equipment like tractors and large tillers to enter from either end. The design concept is pretty simple. Usually, they have one layer of plastic supported by a metal frame, roll-up sides, and no

If considering a glass greenhouse, ensure it has automatic vents at the top. Even in the cooler seasons when seed starting, if the sun is shining, the glass will heat up and trap heat. This is great in some cases but can easily burn up your seedings if it gets too hot.

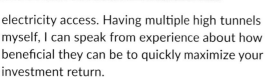

There are many ways you can extend your growing season. If you are not in a place to put up infrastructure, row covers are a great option to extend your season.

able to tolerate the higher humidity inside the tunnel, and be self-fertile due to the lack of pollinators inside. Thankfully there are plenty of greenhouse vegetable varieties that check all these boxes. To save you a bit of research time, here is some more greenhouse variety information, including a list of some of my favorites for high-tunnel growing.

electricity access. Having multiple high tunnels myself, I can speak from experience about how beneficial they can be to quickly maximize your investment return.

If you do decide to grow in a high tunnel, you will need to be selective about the crops you choose. Not only will you need varieties with high yields, they will also need to be disease resistant,

GREENHOUSE VARIETIES FOR HIGH-TUNNEL GROWING

'Grand Marshall' tomatoes

'Excelsior' cucumbers

'Katrina' cucumbers

'Sakura' tomatoes

'Sprinter' bell pepper

'Rebelski' tomatoes

Vegetable types defined

Determinate These varieties are sometimes referred to as "bush" types. They grow to a determined height and then stop growing. They typically mature earlier and all the fruits ripen at the same time.

Grafted A technique that joins parts from two or more plants together so that they grow as a single plant. The upper part (scion) of one variety is joined to the root system (stock) of another. This is done to show improved disease resistance, dwarfing traits, or other benefits.

Gynoecious A variety that only produces female flowers. Production varieties are typically also **parthenocarpic**.

Heritage These are varieties that, much like **heirlooms**, have withstood the test of time and are known for their high quality and flavor. Unlike heirlooms they are not always open pollinated and could be hybrids developed from breeding programs.

Heirlooms These are varieties that have been passed down through generations of farmers and gardeners. They are open-pollinated and are known for their flavorful and unique traits.

Hybrid These are varieties that are the offspring of a cross between two parent plants. They are bred for production and to be disease resistant, high producing, and to grow well in various climates.

Indeterminate Often varieties of vining plants that continue to grow throughout the season. They continuously produce fruit until they've been killed by frost.

Our tomatoes, peppers, and cucumbers are on a similar trellis system and will climb all the way to the top of the pipe at the top. It is a show-stopper every time you step into the tunnel.

Parthenocarpic These are varieties that don't require pollination to set fruit. They are great for growing in protected cultures. Typically they are seedless and produce high yields. The seed is more expensive.

Resistant A variety with a certain amount of resistance to a particular bacterial, viral, or fungal disease.

Pros of high tunnels

- High tunnels are controlled environments. You can control inputs, such as water, fertilizer, temperatures, etc. Because of the ability to create an ideal environment for plants, they often produce higher yields.

- High tunnels promote and maintain healthy soil. Because high tunnels are covered, you aren't experiencing soil and nutrient runoff from excess rainwater and all the nutrients you add are retained, encouraging healthier plant growth.

- High tunnels are generally cheaper than greenhouses because they aren't built on a permanent foundation and do not require any utility hookups for heating and lighting.

Our seed-starting high tunnels have wooden bases at the bottom for an extra level of protection against spring winds and potential tornadoes.

Often when we think of high tunnels, our minds immediately shift to thoughts of growing food through the winter. I certainly know this is what I used to think, but you might be surprised to discover what your high tunnel is really capable of producing when you learn to capitalize on it through each season. Every summer we add shade cloth to the top of our tunnel. We use a 30% light blocking density shade cloth, which allows us to grow throughout the summer without worrying about our tomatoes getting sunscald or our summer greens bolting. The extra shade also makes our tunnel bearable to work in during those hot summer months and allows our plants to continue thriving.

Another great benefit of our high tunnels is our trellising system for tomatoes and cucumbers. We follow a system that is similar to one known as "lower and lean" but ours fits our southern climate a bit better. We do what's called a hard-pruned two-leader system. These trellising options will only need to be put in place with indeterminate tomatoes and vining plants like cucumbers. Indeterminate tomatoes continue growing higher and higher throughout the duration of the growing season, unlike the determinate tomatoes that grow to a genetically determined height and then stop. Indeterminate tomatoes can grow well over

7 feet (2 m) tall and need a proper trellising system to support them. Trellising is also important for keeping low-hanging fruit off the soil surface where it might rot.

LOWER AND LEAN SYSTEM

A lower and lean trellising system is used for intensive growing. Plants are hard pruned to a single branch along the twine, string or guidewire support. As the plant reaches the top of the string, an additional piece is let out from the spool and reattached along the trellis guidewire or conduit causing the plant to be lowered towards the ground and leaning over.

THE 2-LEADER SYSTEM

A 2-leader trellising system is often used for tomato growing. The tops of seedlings are pruned before they are transplanted. This forces the plant to have two main leaders growing upward. Two lengths of twine are attached at the base of the plant extending up to the top of conduit or pipe support. Both leaders are then trained upwards using tomato clips as they grow.

Our high tunnel is a 24 x 50 (7 x 15 m) but they come in various sizes and with assorted bells and whistles. The more accessories you add to your high tunnel, the more your price will increase.

WHAT IS SHADE CLOTH?

Shade cloth is a mesh fabric woven together and designed to block out a certain amount of sunlight. You can choose to use shade cloth with anywhere from 5% to 95% sun protection.

Cons of high tunnels

- One of the biggest disadvantages of high tunnels is that they are permanent and not easily moved.

- Some pests thrive in this protected environment, which can be detrimental to your production.

- Another disadvantage is their size and required inputs. High tunnels take up a lot of space. Plus, the cost of a tunnel is more expensive than some of the other options listed, and you will have added inputs, such as irrigation, soil, and ventilation.

- It can be challenging to control the rising temperatures in a high tunnel. Cooling takes place only through ventilation by opening the sides. Depending on your climate, high tunnels can get very hot.

Greenhouses

A greenhouse, also known as a glasshouse, is a structure with walls and a roof, typically made from translucent materials, like glass, but it can also be made from plastic materials.

If you have been a grower for any length of time, you have probably asked yourself if the investment into a greenhouse is worth it. I remember when I had my first greenhouse built. The excitement it created was electrifying. I had longed for this space for such a long time, and to finally see it being built, brought so much into perspective. I could visualize the growth it would bring to my farm. I imagined starting hundreds of plants to sell at a local farm.

That greenhouse represented what I had waited so long to achieve—growth. You don't have to be at a certain scale to consider a greenhouse; a greenhouse can benefit almost any backyard grower. I learned rather quickly the value far outweighed the cost.

> **You don't have to be at a certain scale to consider a greenhouse; a greenhouse can benefit almost any backyard grower.**

Pros of greenhouses

- Provides a controlled environment, making it optimal for plant production and the ability to grow year-round.

- Aesthetically appealing.

- Prevents larger pests and animals from destroying your crops.

- Rain and wind doesn't have an effect on your plants.

Cons of greenhouses

- With greenhouses there is an initial upfront cost. The cost for a greenhouse can range from a couple hundred to a couple of thousand dollars, depending on the size and how many bells and whistles you want it to have.

- There are ongoing costs. Many plants grown in greenhouses require a specific temperature to be maintained, requiring fans, heaters, and sometimes thermostats—most of these require electricity to run.

- There can be a lack of pollinators. Due to pollinators not having access to your plants, some fruits and vegetables may not produce as much if you don't hand pollinate.

Row cover

Row cover is a transparent or semi-transparent flexible fabric that is placed over your plants as a protective covering. Row cover is typically used for vegetables to help shield against harsh winds, cold temperatures, and even pests. It comes in a range of thicknesses and lengths. Use a sharp knife or scissors to cut the row cover to the length of the area you are trying to cover. Even the lightweight row covers will keep your plants from burning under direct summer sunlight. All weights of row cover will offer pest protection too.

There is no one way to use row cover. How you use it depends on the crops you are growing and the time of year you intend to use the covering. It is a vital tool for the small-scale grower. Row covers give you the opportunity to plant earlier and

extend your season for a very minimal investment. If treated properly, row cover can be reused season after season.

Tiny tip

Immediately clean and store your row covers properly when they are not in use to ensure you get the longest life out of them. View this as an essential task that will help eliminate costs in the long run. Row covers can be cleaned in a large bucket of soapy water, rinsed off, dried in the sun or on the line, and folded for storage in a shed or garage.

Pros of row covers

- Row covers are great for shielding out pests. We would not have such a successful crop of brussels sprouts every year if I weren't able to use row covers to keep the cabbage moths from laying eggs on the plants. Not only does the row cover keep away caterpillars in our garden, but also Japanese beetles, vine borers, and so many other damaging insects.

- Row cover protects plants from the harsh sun.

- Until they become more established, lay row cover over freshly transplanted seedlings in the spring to offer some protection from the sun, rain, and wind.

Cons of row covers

There are not many cons when it comes to investing in row cover for your gardens. It is relatively cheap and can serve many purposes. Here are a couple of things to consider before investing in row cover:

- If you are using row cover throughout your entire growing season, you will need to add supports that are tall enough to allow your plants to continue growing. This will take additional time and cost. I use a 9-gauge wire to create support hoops. I cut it to the

A row cover is an inexpensive option to extend your growing season past the first frost. If you take care of your row cover it will last you for years.

desired length and place the hoops every 5 feet (1.5 m). The closer you place your hoops together, the more support you will have. You can also make these hoops out of PVC pipe or bend electrical conduit into an arch shape. Rebar works, too. Each of these options will be in a different price bracket.

- If you use row cover as a season extender, an early-season jumpstart, or simply for pest protection, you will have to continually remove the row cover to cultivate beds, harvest produce or flowers, and keep an eye on your plants. This can become extremely time-consuming depending on the amount of square footage you have under row cover.

Tiny tip

To save costs on securing your row cover, take a quick walk around your farm and see if you have any materials you can reuse. I have used cinder blocks from other projects, large rocks, and pieces of wood to keep row cover in place. This saved me from having to buy sandbags and sand to secure the cover.

Cold frames

A cold frame is a transparent-roofed enclosure. Its purpose is to protect plants from adverse weather conditions. Typically, cold frames are built low to the ground but slanted at an angle to let more sunlight in. Cold frames utilize solar energy and create a microclimate for your plants.

If you are considering building a cold frame, ensure you are using quality materials that will last through the years.

Pros of cold frames

- Cold frames are cheaper than a greenhouse or high tunnel.

- Cold frames give the opportunity to get a head start on the season.

- They can be used to harden off your plants.

- Cold frames protect against pest and animal attacks.

- In the winter, cold frames can be used for storing overwintering dormant plants.

Cons of cold frames

- Most cold frames are small, so they are not ideal if you are starting or growing a lot of plants.

If your growing climate experiences harsh wind and snow conditions, cold frames will hold up better than row cover.

- Overheating is a possibility if you don't manually lift the lid of your cold frame or have an automatic lid installed.

- Cold frames will not stand the test of time if they are built of flimsy materials.

- Cold frames do not stay warm enough for most tender annuals to survive the winter.

Season Extension Comparison Chart

Structure	Pros	Cons
Caterpillar Tunnel	• year-round growing • easy to build & move • doesn't require level surface • low-cost design • excellent for beginner growers	• aren't as structurally sound • additional cost to replace plastic • difficult to raise (and keep up) interior temperatures during the winter
Low Tunnel	• low-cost investment • easy for one person to assemble & move • overwintering protection for plants • saves space vs. larger options • water conservation	• clearance is not tall enough to walk in and stand upright • manually need to lift sides if necessary to vent • may not withstand harsh conditions
High Tunnel	• year-round growing in-ground • moderate temperature control • heated by sun and vented manually • reduced pest and disease • increased yield	• moderate startup cost • too large for home gardeners with small spaces • not easily moved • initial field prep required (level surface) • replace plastic every 3-5 years
Greenhouse	• season extension year-round • precise temperature control • heated and vented automatically • transplant or in-ground production • good pest and animal exclusion	• most expensive • building experience needed • frequent monitoring required • will be ongoing energy cost & repairs
Row Cover	• affordable option • lightweight and easy to put on by yourself • keeps out pests • allows you to plant earlier in the season because the row cover warms the soil	• doesn't work well if you have harsh winters with heavy snow loads • you have to continuously remove the row cover to cultivate and harvest • Pests could become trapped under the row cover and do significant damage to plants • chance of increased disease due to the high humidity levels and lack of air flow
Cold Frame	• season extension in winter • DIY project • good for hardening off plants • overwinter growing in cold climates • inexpensive	• limited space • can overheat quickly in hot climates • used only in the winter

Before investing in a greenhouse, calculate your needs. How many seeds do you intend to start? Will you be seeding any plants and need additional space? If you are just growing food on a small scale for your family, you can build a small greenhouse to begin with and expand later if you outgrow it.

Hopefully, you are feeling encouraged and confident about which of these structures will be the right choice for your tiny but mighty farm. They are all viable options for season extension and growth but vary in price and commitment. Now that you know your options, what could these structures do for your farm's potential sales growth?

The next chapter covers advice for growing in your community and information about what it's like to take your products to consumers. If this is something you've been eager to learn, then read on. It is going to be a lot of information, but it's all meant to set you up with the right tools to feel confident spreading your wings and transitioning into the new and exciting world of selling the bounty of your harvests.

Planning for community events is a
highlight for me every year.

CHAPTER NINE

Growing for Community: Turning Your Tiny Farm into a Business

When I think about community, so many things come to mind. I have built my business and farm with the concept that community is the driving factor behind everything we do. In this chapter, I am going to not only discuss how to sell your products but also discuss why cultivating a community-driven farm is valuable for your success. We discussed in chapter 1 why small farms are vital to the growth of your local economy, and now we will walk through how to achieve that.

First things first.

☼

Know your market and their needs

Before you can start selling your produce or flowers to consumers, you have to first know your market and understand your future customers' needs. There are a few ways to go about this. You can send emails and make phone calls, but for me personally, there is something about meeting face to face and shaking hands. This lets your community know you are invested. Carving out time to meet in person with your potential customers to understand what they are looking for product-wise makes a significant impact on your relationship with them in the future. This is not only an excellent opportunity for them to express what their needs are, but it's also a chance for you to share with them how you intend to meet their needs and serve them on a weekly basis.

I like to think of it like this: if I were the consumer, how would I appreciate someone approaching me? It's always beneficial to put yourself in their shoes and find ways to make yourself and your products stand out. Anytime I meet with a potential restaurant, I bring them various types of produce to try and I ask for feedback on how I could better serve them. This approach has been welcomed by many, and it's how I put in the groundwork for what I hoped would be a long-lasting relationship.

> "Community is much more than belonging to something; it's about doing something together that makes belonging matter."
> —Brian Solis

Find your niche

When searching for your customers, it's also important to remember what you enjoy growing. This may seem like common knowledge, but I speak from experience. You can quickly get yourself in some tricky situations. I've heard the term "niche down" but never fully understood it until I learned what it meant the hard way. When I first started dreaming of selling my produce, I was so excited about the potential. I could envision having our tomatoes listed on local restaurant menus or lined on the shelf at the farmstand. I would have grown anything someone asked me to if I had been given the opportunity. While being eager is a great attribute to have, it left me agreeing to grow things that:

1. I didn't enjoy growing.
2. I didn't know how to grow.

Here's where the tricky situation comes into play. When we don't niche down and we instead agree to grow anything anyone asks for, we may not be skilled enough to grow those things. When this happens, you cannot confidently approach your consumers with top-notch produce or flowers because you are still learning how to grow them efficiently. When niching down, ask yourself: what do I enjoy growing? If you despise how prickly those okra plants can be and dread harvesting them, then it won't be the best idea to agree to supply okra to several restaurants. If you struggle with back issues, choosing to grow and harvest lettuce several times a week will get old really fast. Start with what you enjoy growing and what you confidently know how to grow. This will be crucial in taking the next steps.

Another consideration would be your time availability. When searching for new consumers and new ways to sell products, consider how much time you have to invest. Some may be pursuing this as a full-time job and have more time to spend planning, planting, and harvesting. Others may be doing this as a side job while still maintaining a job off-farm. If that's you, your time may be more limited. Remember to set realistic goals and don't overwhelm yourself. Some crops will be more time-consuming, while for others, you can have a more hands-off approach. Choose the varieties you grow wisely.

Run a cost analysis

Once you've established the consumer, your marketing, and the products you intend to sell, you can work on your cost analysis. Knowing the price point you need to be at for your farm's products is extremely important. Let's back up some. Money is such a hard thing to navigate, especially if you are a new grower trying to take your products to market. I know it can be easy to agree to any price because it is better than the price you would have received without having any customers but be mindful not to sell yourself short. Know the market and prices before ever trying to sell the product. Throughout this entire book, I have expressed the value in

writing things down, well, here we are close to the end, and I'm sticking with it. Grab your paper and start writing.

Start by listing:

1. The crops you nailed down growing.

2. What kind of consumers are you planning to approach?

3. How much income do you need these crops to generate?

4. The current market price for the crops you are growing.

If you are growing organically or certified naturally grown, you will be able to get more for your products than if growing them conventionally.

If you are unaware of how to go about determining the market price for a product, head to the local farmer's market or a nearby grocery store. Most of my camera roll on my phone in the early years of farming were images of the aisles at the grocery store. That is always a good place to start. Another question to ask yourself is how you plan on selling: wholesale or retail? This makes a significant difference in the price brackets. Many of the customers I sold to also told me the price they were willing to pay, which allowed me to compare to other markets and better understand my costs.

COST ANALYSIS PLANNING

When planning your cost analysis, it is crucial to account for potential crop losses or any hiccups that may occur throughout the growing season. If you plan for this earlier in the season, you will be able to make up revenue in certain areas for those "just in case" situations that may arise.

Know your price

My biggest piece of advice to better understand your ideal pricing position is to create a spreadsheet. I recommend creating a spreadsheet for most things on your farm, including crop plans, planting dates, and crop yields, just to name a few. It takes some upfront time, but the records you have on hand for years to come will make it well worth doing it.

One of the spreadsheets I encourage you to have is a cost analysis. Here is an example of one I use. It lists how much I spent on seed, soil, containers, and other various materials used to grow a crop. Knowing your inputs will allow you to determine the price you need to get in order to sell your products at a profit. Nothing is worse than not being organized and underestimating your startup investment. Again, every word written in this book comes from hard lessons learned.

If you sell your products for less than it costs you to produce them, your business will not thrive and grow. This is another reason why it's important to implement those efficient systems to cut down on the cost of your labor. Knowing your numbers and how to become profitable is crucial. I know many growers who work seven days a week, all hours of the day, and aren't even breaking even by the end of the season. This is not sustainable and will result in burnout.

However, when thinking through your price point, consider more than the cost of inputs; consider what is sustainable for you and your family. Start small, and don't be afraid to stay small. After all, I did write an entire book on its value.

Start small, and don't be afraid to stay small.

Finding customers

After determining your inputs, start evaluating which market you are going to pursue. Will you market your products direct to the consumer? Will

If you grow a crop with the intention to sell it but then have difficulty doing so, you need to know that it is a crop you and your family will enjoy. When you do this, you are ensuring that no food is wasted even if it can't be sold.

you sell wholesale or retail? In a moment, I will list what each of these markets entail. It is an overview only, but before we elaborate on each market, I want to walk through what it looks like to acquire each customer, and also how to keep them as regular customers.

Earlier, I mentioned requesting an in-person meeting and bringing examples or samples of the products you want to sell. Another great option that works well, not only for in-person meetings but to keep on hand to give out to new potential customers, is a farm brochure. Now, before you get the idea that this is an old-fashioned way of marketing, hear me out. If you think outside of the box and get creative, you can also use this brochure to sell your story. I prefer to have mine with pictures of my family and farm so that there is a face behind the product.

I have piles of notebooks for various things. One is to track my costs and is completely broken down by crop. Another is for customers and what they purchase weekly along with invoices sent. I also keep a notebook of general notes; crops that did well, things I won't grow again, etc. If I did not keep detailed notes of these things I would not remember all the important details throughout the growing season.

Hit the highlights. What does your farm specialize in? What story do you want to tell? There are many ways to get your story and your farm in front of consumers, but the best way to sell your produce isn't by marketing your products at all. It's by selling your story.

If I could offer one piece of advice to the small-scale grower, it would be this: quality matters. Yes, quality products are included in this statement, but so are quality relationships. Do not overlook this. It is of the utmost importance. You will feel the need to "sell" your business to consumers around you, but when you produce high-quality crops, it eliminates the need to do so. When you grow in harmony with nature, it shows. Most customers are more than happy to be buying produce grown and harvested a few miles down the road rather than having truckloads shipped in from who knows where. They understand the impact it has on the local economy. Granted, I wouldn't try to sell your product to fast food chains, as their values will likely not line up with yours, but many farm-to-

table restaurants and mom and pop shops love supporting the local grower. It is important to share your values with your customers as well. This paints a clear picture of the farmer they are agreeing to partner with and builds a quality relationship.

> **Most customers are more than happy to be buying produce grown and harvested a few miles down the road rather than having truckloads shipped in from who knows where. They understand the impact it has on the local economy.**

Let's talk next about how to determine what market to pursue.

Every setback is an opportunity to grow and learn as a gardener. Stay encouraged through this process of learning and transitioning into selling your produce.

Direct to the consumer

Direct to the consumer is a business model of selling products directly to customers and bypassing the use of a third party. There are multiple ways to do this.

Farmer's market

The farmer's market is a tried and true way to sell your produce. While this is not an avenue I personally ventured into, it has been a good way for farmers to sell their produce for years. I may not have explored selling at a farmer's market, but I set up sales booths at several conferences every year and display my products much like I would if I were selling at a farmer's market. Ray Tyler of Rose Creek Farm's motto is "Pile it high, and watch it fly."

After taking his master class, I fell in love with how he sold at the farmer's market. His booth was always full. You'll never find an empty spot anywhere. By marketing this abundance approach, he would sell out every Saturday.

A few things to take into account when setting up your market booth.

- Is it easy to shop?

- Is everything labeled with variety and price?

- Is your booth clean and orderly?

- What is your plan for customer service? Having more than just yourself available to work the booth and answer questions, pack up products,

and be a friendly face is essential. This goes back to how you would want to be treated as a customer.

If you decide to sell at the farmer's market, you will need to figure out what sets you apart from the other booths. It may be growing crops other farmers don't or taking personal requests. By making yourself available, you will build up your rapport with the community, and they will see you as a reliable source for their food needs.

There will likely be vendor fees you'll have to pay to sell your product at the farmer's market and possibly licensing fees (if your state requires them). Farmer's markets are a great option to sell your products at a retail price.

CSA shares

When I first learned about CSA shares, I fell in love with the idea. I knew spending every Saturday at the farmer's market would not work for my growing family, but I still wanted a viable option to sell products to consumers. As I started to learn more, I knew CSA would be an avenue I explored. You may already be familiar with what a CSA share looks like, or perhaps you are like me and your community lacked any CSA's. After learning what it was, I started to research it and found no surrounding farms offering this. For someone wanting to scale their business, this was excellent news. I wouldn't be competing with other CSA shares and growers. It would be an opportunity to be a pillar for my community with this style of selling produce.

WHAT IS A CSA?

CSA stands for community-supported agriculture. Otherwise known as "crop sharing," it is a way to connect farmers and consumers. Most growers offer subscription boxes that ensure the consumer receives a share of the crops the farm produces for a specific price and period of time.

I could spend all of this chapter telling you the stories of when I started selling produce. Some stories I laugh about, others I cry about. It was a season of major growing pains. I was eager but had no idea what I was doing. I started selling produce when we lived on a 1-acre (.4 ha) plot, and only used a small portion of that space for food production. I knew I wouldn't be able to offer a lot of produce consistently because I would still be learning as a grower, so I took the idea of a CSA and added my own spin to it.

I worked out religiously at my local CrossFit gym, and I can tell you, that is one group of people who care about what they eat and care even more about supporting their community. I started talking to the members and the owners, expressing my desire to sell food shares weekly. Much to my surprise, the response was tremendous. I posted weekly what was available, took online payments, and made deliveries twice a week. Because I didn't promise any specific crops, it allowed flexibility as I was learning to maximize my yields

If you are in a similar stage of growing your farm, you might think differently about CSA shares. Instead of promising crops for a certain length of time, tailor your CSA to what you can offer from week to week. Most CSA programs limit the number of customers they take per share season. This allows you to build a very loyal customer base and adjust to the needs of your customer.

The first encounter someone has with you or your booth staff will unfortunately be the only one they remember. Come with a clear mind and ready to serve your community.

to take a walk around the farm, look at crops, and connect with the grower, then pick up their weekly produce.

It doesn't have to be this romantic for you, but the idea of allowing the customer to see where their food is grown is a powerful motivator to connect the community with the grower.

For some, it may not look like farm tours or conversations over a glass of fresh lemonade. It may be simpler and hands off, and that is okay. When choosing how to model your farm and your business, you are in control of how you decide to sell and move products.

Some farmers may place a stand at the front of their farm and have it lined with fresh produce, flowers, and additional add-ons to purchase. Some may designate Tuesdays from 11-3 pm for farm pick up. Regardless of the model you choose, farm pick up is a beautiful way to sell your produce and reduce costs. Not only can you sell at a retail price, but you also eliminate money and time spent on making deliveries.

Along with farm pick-up options is the possibility to do pick-your-own (you-picks). We will be implementing this in the summer for dahlias on our farm. It is a great way to get exposure. You receive feedback from the community as they come harvest what they want to buy, and you can sell your products for a more than fair amount.

Tiny tip

If implementing a CSA share program on your farm, include "add-ons" for additional income. This can be specialty products like microgreens, salad mixes, bread, jams, eggs, flowers, and so on. The options for additional add-ons are endless. This is an efficient way to capitalize on existing customers and to grow your income.

Farm pick up

Doesn't "farm pick up" sound dreamy? I can imagine myself harvesting vegetables in the garden while cars pull up the drive. Customers get out

Do a trial run for you-picks with your friends and family. Encourage them to come out and try this method of buying at a discounted rate in hopes that they will spread the word and bring in future customers. You will be able to charge more than what the produce or flowers are worth because you are not just selling the product. You are providing an experience.

Wholesale options

Wholesale customers are consumers who are buying a large number of products for a discounted rate so they can, in turn, sell them and make a profit themselves. There are several different wholesaling opportunities for a tiny but mighty farm.

Restaurants

Restaurants are always wholesale accounts. Your farm would need to follow an intensive growing approach to be able to supply the full demands of most restaurants. If you are implementing a market garden approach, this is a great way to move a lot of product because you can grow a bulk amount of a few varieties and deliver them to the same place on the same days each week. One way to get a restaurant's attention is to grow unique varieties; this can prove lucrative for both the grower and retailer.

Growing diverse food will make you a well-rounded grower, and many chefs will appreciate your willingness to try new varieties.

speak to the person in charge of ordering produce. There may be some smaller grocery stores that will give you the time of day, but if you live in a bigger city, this will be more difficult. In my experience, unless you are already a reputable intensive market gardener, this is the most challenging market and the least profitable approach to selling to consumers.

Co-ops and food hubs

If you have a food co-op or food hub in your area, you understand firsthand how vital they are to the local community. We have a food hub called "The Farmstand," which I will highlight later in this chapter. But, essentially, what Kim, the owner of the farmstand, has created is a locally sourced grocery store. Everything she sells is grown or produced by a local farmer, artisan, or crafter. She sells anything from fresh bouquets and produce to value-added products. You name it, The Farmstand carries it.

APPEALING TO CHEFS

If you are trying to get the attention of local chefs, I highly encourage you to ask them to come tour your farm. They can get a feel for what and how you grow, the systems you have in place, what your produce looks like, and other growing practices that could be beneficial for your farm's growth.

Grocery stores

Selling to the grocery store market often proves a bit more challenging. Most of the time, you are not able to walk into a grocery store and ask to

---

WHAT IS A FOOD HUB?

A food hub, as defined by the USDA, is "a centrally located facility with a business management structure facilitating the aggregation, storage, processing, distributions, and/or marketing of locally/regionally produced food products."

Getting your products into a food hub like this could be beneficial for your growth and exposure as a farm. The pricing will be based on wholesale prices, but in my experience, I've always been paid more from food hubs than wholesalers like restaurants. The only way to know if these co-ops and food hubs exist is to ask around. Food hubs are familiar with buying from multiple producers and typically have a price list they can give you, which simplifies your pricing if you are new to selling this way.

Value-added processors

Selling to restaurants and grocery stores isn't the only way to move product on a wholesale level. Value-added processors are another contender for a customer market. With these accounts, you sell them the raw product (whether vegetables or flowers), and then they use the raw products to make a retail product, such as jams, pickles, elderberry syrup, flower bouquets, or other value-added item. This is a great way to embark on a partnership with other local producers in your area. We grow ginger for a company that makes elderberry syrup and sells to multiple online stores and local brick-and-mortar stores. This is fun for me as a grower because we had a hand in developing a product that we otherwise could not and would not have produced on our own.

I understand how much information I just threw at you, but if you are trying to transition into selling your products, I genuinely believe what this chapter has to teach you will catapult you and your farm into success.

More important factors

Before we shine a light on farmer highlights, let's look over a few more components to consider before selling to a market.

Land size

This book is proof you don't need a lot of land to grow a significant amount of food. But you will need to figure out how to focus on those systems we discussed in chapter 7. The more organized your systems, the easier it will be for you to sell and market your produce.

Soil

We talked in-depth about soil in chapter 4, so hopefully, you understand how vital it is to the health of your tiny but mighty farm. Thankfully though, quality soil can be established on any property. It all depends on what you start with and how much work you are willing to put back into it. The quality of crops you have to sell will be dependent on how much time you put back into building and maintaining your soil quality.

Location

This can be important for a few reasons. If you live way out in the country and only plan to sell through farm pick-up, you may become discouraged by a lack of growth. On the contrary, if you live in an urban setting, you could max out farm pick-ups because of how convenient it is.

Labor

If you are the only one working on your farm to grow, cultivate, harvest, and deliver, be considerate of how many crops you agree to produce and how many markets you are selling to.

Be genuine

Last but certainly not least, when learning to sell your products and advertise your farm, be genuine. Authentically share your story and extend an invitation for your customers to come to see your farm and ask questions. Be friendly, paint an honest picture, and share the ups and downs that come with farming. Most importantly, stay encouraged. It may take you months to land your first customer, and that's okay. Keep showing up, be willing to put in the work required, and you will reap the benefit.

CULTIVATING COMMUNITY

In a culture where everything is about competition, I encourage you to build relationships with the mindset that community trumps competition. Think of creative ways to link arms with surrounding farms, neighbors, and friends to cultivate a community-driven atmosphere that others will want to model. This will require you to be intentional with your relationships and work through hard situations. Still, there has never been a relationship I've pursued that has not proven to be valuable to all parties involved. We truly are better together than we are apart.

Farmer highlights

I couldn't write a book about my community values without highlighting some of the farmers who have been pillars for my community but also who have mentored and encouraged me through the years. These three growers demonstrate (on different scales) all the things we've discussed in this chapter. I hope their stories will motivate and encourage you to keep working toward your goals and always remember how your tiny farm can produce a mighty impact.

THE BACKYARD GROWER

Name: Sean and Melanie Pessarra of Mindful Farmer

Size: ¼ acre (.1 ha)

Website: mindfulfarmerarkansas.com

Location: Central Arkansas

Sean and Melanie Pessarra own Mindful Farmer and Mindful Kitchen. It is a family-owned business in Central Arkansas that's on a mission to equip gardeners, homesteaders, and farmers with the tools and skills they need to best steward their land, homes, and businesses.

Sean specializes in crafting greenhouses and tools for production purposes. He uses his backyard chickens rotationally to prepare his flower fields. He and his wife also run a small cut-flower

Cut flowers have become the most profitable crop per square foot. They benefit significantly from being grown under the protection of a tunnel. It increases stem length and bloom quality.

business in their backyard. They are growing on a thousand square feet (93 sq. m), half of which is high tunnel production. He hopes that more people will labor using sustainable agricultural practices and thoughtful consideration to improve their communities, land, homes, and health—despite the small area they have to utilize.

Because Sean spends most of his summers building custom greenhouses and consulting for local farmers, he puts a significant emphasis on forcing bulbs like tulips, ranunculus, and anemones, throughout the winter. Doing this allows him to max out his small space, bring in profit throughout the winter and early spring, and free up his summers.

Mindful Farm began as a small, family-owned beekeeping operation in Texas. A few years later, they had the opportunity to step full time into sustainable agriculture in Central Arkansas. With Sean's background in ecology and experience on the farm and Melanie's background in nutrition, they saw the impact conventional farming practices have had on their soil, food, human health, and ultimately the world's health. Their vision has

continued to grow beyond bees and into so much more.

As they've grown their business, they certainly have faced some challenges. They envision creating a locally driven farm and want to impact their community by offering a thousand-share CSA. To do this, they must first bring awareness of the importance of locally grown food to the community while also finding access to land. Lack of land access has proved the most challenging of all.

Despite some of the setbacks they have encountered, they have successfully ventured into a profitable cut-flower business that requires low maintenance due to their implementation of efficient systems early on.

They work closely with us at Whispering Willow Farm, share space and resources, and sell wholesale to Bell Urban Farm.

One of their greatest joys is to work alongside other farmers and growers of any scale to provide organic food and locally grown flowers to their communities.

THE URBAN FARMER

Urban farming or gardening is the practice of growing and producing food or flowers in or around urban areas. Urban farms are extremely valuable for introducing local organic food and farm values in a city setting that may be more convenient to most people.

Name: Kim Doughty-McCannon & Zack McCannon of Bell Urban Farm

Size: 1.5 acre (.6 ha)

Website: bellurbanfarm.com

Location: Conway, Arkansas

Here is an example of an arrangement you can find freshly made at the farmstand. Kim is holding 'Sahara' rudbeckia, 'Green Mist' ammi, statice, and basil.

Kim Doughty-McCannon and Zack McCannon started Bell Urban Farm in 2017 on a small patch of lawn in front of their home. Today their farm is an acre and a half, with 1/2 an acre in production for cut flowers. They have two small children, an extensive backyard garden, two greenhouses, honey bees, chickens, and a local food grocery store, The Farmstand, which opened in 2020. Bell Urban Farm is a Certified Naturally Grown growing operation.

They started their farm growing a mix of veggies and flowers and selling exclusively at the farmer's market. Each week they would set up their booth and notice the lack of access their community had to organically grown food and flowers. That is what sparked the idea of The Farmstand. They saw a need for their community and immediately started working to fill that need.

Kim and her husband lived next door while they bought The Farmstand (the house next to them) and started renovating. However, building The Farmstand did not come without its own set of challenges. The Farmstand had to be rezoned from residential to commercial and that involved many city fees, applications, and guidelines.

They were able to see their dream become a reality through crowdsourcing and their community members joining arms with them to make their vision come to life.

For the past five years, they have hosted an event, the "Spring Plant Sale," and spent countless hours orchestrating the collaboration of surrounding farmers, local artisans, restaurants, and many others to provide their community with access to organic, locally grown plants, food, and music.

Kim has been a pillar for her community and, through partnerships with other growers, has offered exposure and revenue, helping them grow their business through the community.

Her biggest advice to any new farmer is to seek out mentorship from other locals. Finding someone close to you and within the same region and growing zone significantly impacts how successful your endeavors can be.

She also recommends seeking out several different revenue streams and markets to ensure you move your product throughout varying seasons. Despite the ebb and flow of being a farmer, business owner, wife, and mother, Kim finds great joy in her small farm's impact on hundreds of people.

THE FULL-TIME FARMER

Name: Brandon and Cat Gordon of Five Acre Farm

Size: 1 acre (.4 ha) or more

Website: 5acrefarmsar.com

Location: Bradford, Arkansas

Brandon and Cat Gordon are located in Bradford, Arkansas. Brandon founded Five Acre Farm in 2009. There are 5 acres (2 ha), but only 1.5 acres (.6 ha) of that is designated for production. They specialize in storage crops and lettuce, which allows them to have a more consistent supply throughout the year for their customers.

Brandon and his wife run a no-till–style farm that is sustainable and human-powered. They demonstrate both in-ground growing and have around 16,000 square feet (1,486 sq. m) of high tunnel production, which helps them achieve their goal of providing food year-round.

Brandon started the farm on a small piece of ground. Shortly after, Cat, now his wife, joined him. They've since welcomed three children, more land, and greenhouses to expand their no-till operation. Along with the root crops they grow, they also have tomatoes, peppers, cucumbers, and various brassicas for particular markets.

They currently sell to natural grocers, restaurants, and food hubs. Their farm offers a small share CSA through their online store as well.

Brandon is modeling a market garden approach and is demonstrating how efficiently you can grow food on a small acreage to provide goods to various markets.

One challenge he has faced each year, and increasingly so, is the weather. We all know it can be unpredictable, but Brandon says he has seen an increase with each season of heavy rainfall, drought, extreme heat, and deep freezes. The weather can be tough to navigate when you are farming full time.

When Brandon focused on his lettuce production, he started seeing significant successes

Brandon and Cat have made their farming a family affair and welcome their children to help. Pictured here is 'Black Magic' kale.

with his farm. By specializing in lettuce and roots crops, he was able to have a more predictable income week to week throughout the year. Through that success, he was also able to expand his walk-in cooler spaces, which allowed him to do even more in these areas.

His advice for new farmers is to start small; very small. He even recommends having a job off the farm until you've become familiar and confident in your endeavors. He encourages other aspiring farmers to set up at the farmer's market to get an idea of the demand in their area and work towards catering to that demand. Another piece of advice he has for small farmers is to not go into debt until they have a clear idea and vision, and ultimately until they are confident enough to deliver on that.

Brandon has been an anchor for his community through both his crops and the efficient ways he grows them. By supplying to multiple markets, he has a significant impact on his local economy and the food his community members have access to buying.

Community events

I love community events. Well, let me rephrase that. I love community events that revolve around gardening and growing food. Despite what some might know about me, I can easily transition into the biggest introvert you've ever met if you put me in a setting with a large crowd, talking about things I know nothing about. I don't find joy in talking about the latest show on television or the newest song to hit the radio, but I sure can tell you all the new varieties in my favorite seed catalog or the new heirloom vegetables I'm growing on the farm.

If you are reading this and that resonates with you, you'll enjoy this next part. Bell Urban Farm has hosted a plant sale every year for the past five years. I have been a vendor for four years, and it is one event I look forward to each year. Imagine several different farmers growing the highest quality organic plant starts, live music, local food and coffee, and community members that show up from surrounding cities. It is a day of pure bliss. The gates are lined with people waiting for us to open; it's chaotic and beautiful. I am always blown away by the number of people who show up to this sale every year. They are the reason I do events like this. They speak words of encouragement, rave about the varieties we offer, and share the love between all the farmers. For me, this is what the community represents—showing up and supporting one another. I will forever be grateful to Bell Urban Farm of Conway, Arkansas, for putting in so much time and attention to see the success of not only this event, but the success of each grower involved. Kim, thank you!

Another excellent way to become involved with community events is to attend or host farm-to-table dinners. This is one thing that is front and center on my dream board. I've had visions of what it would look like to host farm-to-table events on our farm. When I think about events like this, I envision partnering with other farmers and highlighting what they specialize in. For some, it's meat. For others, it's produce or flowers. But for all of us, it represents the community.

Kim is the owner and founder of Bell Urban Farm and has been a wonderful mentor to me throughout the years.

If your community lacks events like this, I'm going to put some pressure on you. Not the kind of pressure that forces you into something, but the kind that pushes you out of your comfort zone and into a place of growth and gratitude. I used to think I'd never be able to pull off events like this on my farm, but now, as I am in the planning phase for our first on-farm workshop, I realize how vital this leap of faith was to me and will be to others.

I've always said if you see the need, work toward filling the gap. My hope is that after this chapter you feel equipped to start marketing and selling your produce either on a small scale or a large scale. Let the backyard grower, urban farmer, and full-time farmer be a testament of what you can accomplish on a small plot of land. If you try one of these approaches to sell your produce and it doesn't suit your needs, try another until you've found the avenue that aligns with your values. And let's not forget the most important thing. No matter what scale you are growing, you always have a seat at the table.

I've always said if you see the need, work toward filling the gap.

I could not imagine growing food for my family and my community if I wasn't able to tangibly see the positive impact it was having. Make time for opportunities that will grow you as farmer.

Conclusion: Keep Dreamin'

Hey Friends,

You made it! I'm so proud of you!

You now have the information you need to embark on a beautiful journey to create the farm of your dreams and provide abundant food for your family right outside your own back door. The best part is it doesn't matter if you are growing on a ¼ of an acre (.1 ha) or a couple. The words in this book still apply. One of my favorite things about learning is the ability it gives me to teach others. I hope this book encourages you to cultivate community and pour your knowledge into others starting on their farm journey. When you walk into your gardens, my hope is that this book is in hand, that you jot notes down on every page, reference the charts, and use it as a tool as you learn to grow and scale your own tiny but mighty farm.

"The capacity to learn is a gift; the ability to learn is a skill; the willingness to learn is a choice."

—Brian Herbert

Dreaming freely is a privilege I will never take for granted. I hope you shoot for the moon every time, and even if you land in the stars, it beats sitting on the ground.

Cultivate dreams

In this book, we covered everything from the values small farms play in our local economy, defining your purpose, gardening styles, soil health, and so much more. While I want you to remember all the valuable information packed in these pages, the main takeaway I want you to remember is that you are worthy of every dream you write down on paper. When casting a vision for your farm, cast far and wide. Dream of the impossible and live with anticipation for when that day comes to pass. I can teach you all the benefits of high tunnels and greenhouses and why community is essential for growth, but what I can't teach you is how to dream. Only you can do that for yourself.

I get it; sometimes dreaming is hard. Not all of us typically spend much awake time dreaming, but I earnestly believe that whether you are a realist or a romantic dreamer, there is value in seeing your worth and what you are capable of achieving. It's not always about the long-term goal. It's more about how we grow as gardeners getting there. What were the lessons you learned? What community did you cultivate? What systems did you put in place to push through the setbacks?

You are worthy of every dream you write down on paper.

Keep records of all your growth, failures, and successes. It has been a joy for me to look back on photographs and documents of how my tiny backyard garden grew into a beautiful, efficient farm. Having these visual reminders motivates us, and it is an heirloom to pass on to the next generation as a testament to the hard work, sweat, and tears poured into stewarding what we've been given.

For me, it's one thing to encourage someone, and it's another thing to walk alongside them and hold them accountable. Since I can't physically do that for each of you, this book is the next best thing. It's a call to action. Yeah, you heard me right. I am going to hold you accountable to dream wild. Grab that pen, and let's finish this book off right. Use the blank space to dream big, cultivate your vision, and believe the impossible is within reach. To the new gardener learning and the experienced gardener growing, I raise my glass.

I'll talk to you soon.

Jill

Acknowledgments

This book is one of my biggest accomplishments. I truly felt as though it's been my baby for the last several months, and now it's out in the world, in the hands of hopeful and motivated growers of all kinds. I couldn't sit here and write this, though, without the village I've had supporting me through this season.

First and foremost I thank the Lord for equipping me with the ability to write this book and use these words as encouragement to many far and wide. I pray for each person who opens the pages to be planted with a seed of hope and joy, and that your love multiples. May my life be a reflection of who you are.

To my sweet husband, Nathan. I cannot express how much I love you. Every late night you spent lying beside me as I overwhelmed you with yet another ambitious dream. You never doubted what I was capable of and encouraged me every step of the way. Thank you for being dedicated to our family and pursuing my dreams as though they were yours. You will forever be my strong rock.

To my sweet daughters, Paityn, Charlee, and Ivy June. Each of you is the reason I wake up every day and put one foot in front of the other. I want you each to pursue your dreams wildly, and

know that there is nothing you can't accomplish. Let this book be a testament that if your Mama can do this, how much greater impact will each of you have! Don't be fearful of your dreams, but tackle them head-on in anticipation of the impact they will have.

To my dear friend Sean, I will never be able to repay you for all the knowledge you've poured into me. You believed in me when I doubted myself, encouraged me when I needed it the most, and helped me plant the seeds of my future. For that, I thank you.

To my friend and photographer, Bailee. Who would have known as 16-year-olds riding horses in the pasture that we would be on this journey together? Thank you for being my friend through the years and bringing my farm to life through the photographs in this book.

To my friend Jessica who introduced an heirloom seed to me. You showed me that dreaming was worthwhile and that I was filled with purpose. Thank you for pouring into a farm that I now steward and for reminding me to always stop and smell the basil.

To my family and friends who encouraged me to plant another seed and sow into another person.

To the ones who understand the garden's value and reminisce with me as we embrace her beauty.

To our social media family, thank you for always believing in any project I've set my hands to do. We would not have been able to see this come to pass without each and every one of you partnering with us. We've learned as much from you as you have from us. You've seen me build my dreams from the ground up and walked with our family through many seasons. Thank you for allowing us to share our lives with you as we continue to grow on this beautiful journey.

To Drew and Jeremy, referred to as the Dream Team. I could not imagine having greater individuals on our team helping us pursue and do the hard things. Jeremey—thank you for being invested in the vision of Whispering Willow Farm and your continuous work to help us grow. To Drew, thank you for the countless hours you've spent helping me with this book, from reading chapters over and over again to your upbeat spirit always motivating me to keep moving forward.

To Cool Springs Press and The Quarto Group for believing I could write this book, and my editor Jessica for offering kind words as we navigated the ups and downs of book edits and revisions. A thank you will never be enough to the multiple hands involved in our team bringing this project to life.

To our community. I shout from the rooftops my appreciation for every one of you. You all played a role in where we are today, from supporting us in the beginning stages, buying produce in the gym parking lot, or supplying your restaurants with farm-fresh produce. Every plant that was purchased in the spring to flowers and vegetables in the summer. You are the backbone of why we farm and garden and the motivation that keeps us moving forward. We are only as valuable as the people we pour into, and I couldn't imagine a greater community to call my own.

About the Author

Jill Ragan lives in Central Arkansas and has over 10 years of experience growing food. She has spent the last 5 years growing her online platform through YouTube, Instagram, and her online home and garden mercantile store. Jill, alongside her husband, Nathan, is community-focused and prioritizes on-farm workshops, and farm events. Jill has aspirations to grow her influence to better serve those around her through online education, books, and hands-on experience. She has worked diligently to turn her passion into a career and is inspiring others that what once was a dream can become a reality.

Notes

Index